USING AN ART TECHNIQUE
TO FACILITATE
LEADERSHIP DEVELOPMENT

touch•stone \ˈtəch-stōn\ *n* (1530) **1 :** a black siliceous stone related to flint and formerly used to test the purity of gold and silver by the streak left on the stone when rubbed by the metal **2 :** a test or criterion for determining the quality or genuineness of a thing.

Reconciling both senses of the dictionary definition, the touchstone is at once a concept *and* an object existing in space and time. It is a lively embodiment and reminder of deeply held ideas and values that can be regarded, reflected upon, shared, and tested.

USING AN ART TECHNIQUE TO FACILITATE LEADERSHIP DEVELOPMENT

Cheryl De Ciantis

Center for Creative Leadership
Greensboro, North Carolina

The Center for Creative Leadership is an international, nonprofit educational institution founded in 1970 to advance the understanding, practice, and development of leadership for the benefit of society worldwide. As a part of this mission, it publishes books and reports that aim to contribute to a general process of inquiry and understanding in which ideas related to leadership are raised, exchanged, and evaluated. The ideas presented in its publications are those of the author or authors.

The Center thanks you for supporting its work through the purchase of this volume. If you have comments, suggestions, or questions about any CCL Press publication, please contact the Director of Publications at the address given below.

Center for Creative Leadership
Post Office Box 26300
Greensboro, North Carolina 27438-6300
336-288-7210 • www.ccl.org

Center for
Creative
Leadership

NORTH AMERICA EUROPE ASIA
www.ccl.org

CCL No. 166

Library of Congress Cataloging-in-Publication Data

De Ciantis, Cheryl.
    Using an art technique to facilitate leadership development / Cheryl De Ciantis.
        p. cm.
    Includes bibliographical references.
    ISBN 1-882197-09-7 [ISBN-13: 978-1-882197-09-5]
    1. Executives—Training of. 2. Management—Study and teaching—Graphic methods. 3. Leadership—Study and teaching—Graphic methods. I. Title.
HD30.4.D4 1995
658.4'07124—dc20
                                                                                        95-31264
                                                                                        CIP

Table of Contents

Acknowledgments

The deepest of heartfelt thanks and the reddest of red roses go to Bob Burnside and Victoria Guthrie for their vision as well as their support and friendship through many years. With full glass a toast of gratitude to Stan Gryskiewicz, who invited me to the Center in the first place. More than I can possibly say to Diane Downey, Bernie Ghiselin, Chris Musselwhite, and Sheila Pinkel for mentoring me, in very diverse areas. The biggest bear hugs to Marcia Horowitz and Martin Wilcox for their encouragement and enthusiasm, every bit as much as for their editing and know-how. The brightest posies to Joanne Ferguson, whose discerning eye helped these pages spring to life. An armful of flowers to Dianne Young, with whom I have enjoyed true synergy. Bouquets to Teresa Amabile, Jay Conger, Maxine Dalton, Bill Drath, Rob Goldberg, Mary Ellen Kranz, Winn Legerton, Rick Morales, Chuck Palus, Sue Rosen, Marjorie Parker, Don Petrie, and Martha Tilyard, for close readings and invaluable suggestions. A "group hug" to LeaderLab® staff, past and present. Warmest thanks to all the Center staff, who have made this a second home for me and a friendly as well as fulfilling place to be. And to all the LeaderLab participants, for their courage, fortitude, openness, and good humor in the face of even the oddest developmental challenges.

Foreword

This report is a new departure for the Center. For the first time we are giving our readers everything needed to understand and try out an innovative developmental technique used in one of our programs. Not just the theory of the touchstone exercise is presented here but the details of doing it—how to set it up, run it, debrief it, right down to how to pack up touchstones for shipping. Case studies of people using the exercise for development make immediate for the reader the range of thoughts and feelings it inspires. We hope this is useful to you in program design and in thinking about new approaches to leadership development.

This report is also a departure because its author combines leadership training and training design with being a practicing artist. At first this may seem like a strange combination. On second thought, it makes obvious sense. Leadership, like art, is an activity that calls on the whole person. Like art, leadership involves the mind, heart, and spirit. Leadership and art are both essentially indefinable, more something we know intuitively when we see it, than something we can specify and codify. Learning how to practice art and learning how to practice leadership both require more than learning skills; they both require developing humanity.

Still, we tend to think of a leadership development program as working on capacities directly related to leadership. So we wonder about an approach that encourages the person to think about who she or he is, who he or she wants to become. Especially, we wonder about doing this by getting people who are non-artists to do art.

In this report you will discover that people have more capacity for art than they give themselves credit for having. You will also discover that people can use art to understand themselves better, and more, to understand their relationships with others. And that as people discover an unsuspected capacity for art, they can also discover unsuspected capacities for leadership.

Bill Drath
Publication Director
September 1995

Preface

The training technique described in this paper, which uses art to help people develop as leaders, has had a long evolution—much of it related directly to my personal experience.

For more than twenty years I have been involved with art in a variety of ways, working and exhibiting as a fine artist in such media as drawing, painting, three-dimensional assemblage, and computer animation and working professionally as an illustrator, theatrical designer, and graphic designer.

In 1984 I began working in the field of leadership, in the management- and organizational-development unit of an international financial corporation. I took part in the design and delivery of customized programs that were grounded in experiential learning methods (participative, hands-on exercises such as role playing).

When many corporations began to look at new ways of stimulating creativity and innovation, it opened an avenue for me to start to think about potential ways to bring artmaking into the intersection of creativity and leadership. I hoped to help leaders *experience* creativity the way an artist does, and that this would provide an access to creative thinking that could go beyond the creative "problem solving" model. In 1987 I started to experiment by helping managers and executives use the computer as an art tool. My reasoning was that the fascination many people had for the technology could be employed to get them to interact with the computer in a way that didn't involve the use of spreadsheets or financial reporting, and that the visual tools then beginning to be offered by PCs would be an enhancement to creativity and communication. I learned two things from this experiment.

First, the interaction with the machine often yielded a machine-like result. This interaction did not seem to lead to enhanced communication with anything but the machine itself. People would wind up talking to each other about how much memory and what programs to use but not about the content of what they wanted to communicate.

Second, in the early days of the personal computer, when relatively few managers were computer literate, I learned that many people who were reluctant to lay hands on the computer keyboard were quite willing to use a pressure-sensitive digital drawing pad. When overlaid with a sheet of plain paper this particular pad could be used with a pencil to make marks that appeared directly on the screen. Managers who would not touch the keyboard sat in childlike delight with pencil in hand, watching their marks appear on the screen like magic. What I interpreted from this was that the discomfort some people felt with the unfamiliar technology was greater than, or took their minds off, the fear of being asked to draw! Perhaps drawing—using their *hands*—might be more naturally available to non-artists than I had previously believed.

Over the next few years I had the opportunity to try out various ways of incorporating artistic activities in many different facilitative contexts, ranging from cross-cultural awareness, to strategic planning, to building shared vision. I learned the value of providing ways for people to temporarily get "outside the box" of their accustomed mode of organizing experience, by taking a detour around the verbal, usually linear, logical structures which can harden into set habits of thinking.

My exposure to Stan Gryskiewicz's use of "excursion" techniques such as "Visual Connections," originated by Horst Geschka (1993), helped my thinking on this point. In this process, participants are asked to put the problem they are working on aside and are shown a series of visual images of, for example, a rock climber, a cluster of grapes, the face of a cat. Free-associated connections are noted and from them further associations are drawn back to the problem at hand, resulting in novel insights.

Taking a temporary excursion away from words and into images can be useful in breaking blocks in thinking. Working with the corporate audit function of a multinational consumer company in their strategic planning process, my colleague Chris Musselwhite and I introduced a nonverbal process of drawing individual, then "blended," visions of the most important thing each participant wanted to see reflected in the organization five years ahead (Musselwhite & De Ciantis, 1993). The group had been struggling with this plan through two previous meetings. After successfully completing the strategic plan in outline the following day, participants said that the visual, nonverbal excursion had been helpful in surmounting the roadblocks they had experienced previously.

Another idea I found useful, based on innovation practices developed by people such as Rolf C. Smith, Jr., of The School for Innovators, was the simple but powerful notion of sending participants away from an idea-generation session with a tangible record of their output. Thus, not only are generated ideas preserved for further development, but the energy and enthusiasm of the idea-generation session is conserved so that it can radiate beyond the circle of the original participant group. The sense of ownership seems also to be enhanced by seeing the fruits of your labor in front of you.

I put this to use in codesigning a series of computer-facilitated work-shops and symposia within a global financial institution, whose audit division was attempting to institutionalize an innovative way of looking at risk. In meetings ranging from 12 to 250 participants, ideas generated in session were refined into vision statements and action plans which were sent home with the participants in the form of a formal-looking and persuasive document. This permitted participants to "sell" the ideas within their segments of the organi-zation immediately upon their return, which helped to generate customized action plans within a short period of time.

Many of my ideas about leadership development and art merged when, in 1990, I was invited to participate in the development of a new program at the Center, to be called *LeaderLab.*® The designers of this program, Robert Burnside, Victoria Guthrie, and Leonard Sayles, envisioned a holistic ap-proach to leadership development that could confront emerging leadership demands. They were looking for methods that would help managers look at the challenge of their unique situation from a new perspective. The whole human form was chosen as a visual metaphor for the effective leader: one who can engage not only the *head*, or intellect (which they felt has been the sole focus of most leadership training), but also the *heart*, or emotions, and ultimately the *feet*, that is, achieving more effective actions as a result.

Bob, Victoria, and I began to talk about what participants might need in order to "put it all together" from the perspective of the action-based focus of the program. How could they synthesize their various experiences, the con-tent they had explored, the feedback they received, the visions they had created, and the resultant action plans they had formed, and take them back to work?

What began to become clear was the idea that I could make it possible for each participant to create something for him or herself, an object that would be personal and symbolic of the program experience. Better than a generic gift marking the experience and then left to gather dust, this could become something that would actively remind the participant of what is most important in his or her individual sense of purpose in leadership every time he or she looked at it.

A program exercise involving a solitary silent walk in the natural environment surrounding the Center to reflect on leadership issues gave rise to the idea that perhaps materials from nature could be used by the partici-pants to create an object that would be personally meaningful. By using natural materials, they could begin to reflect back in a concrete way on the week's experiences—maybe they could even use something they had brought back from the walk.

Creating this object would be something anybody could do with little assistance. It could reengage the "heart" after the "head" work of action planning—and the participants could "put feet on it" by representing their intentions and commitment in concrete form. Thus, it could provide a bridge from the program experience back to the work environment, where a participant might choose to put it on his or her desk. Victoria came up with a name for it: the "touchstone."

Introduction

In 1991, the Center launched an action-oriented development program called *LeaderLab*,® which uses, in addition to traditional methods like classroom work and feedback, techniques which help participants experience themselves and others in unaccustomed ways. One of these techniques, called *the touchstone exercise*, asks participants to think about their experiences in the program and, using a variety of materials, create an object that will remind them of what they most want to focus on in their workplaces. They then write about its meaning in a "learning journal," and if they wish, verbally share the story with the whole participant group.

The purpose of this paper is to describe the touchstone exercise and to give examples of its use in a leadership development program. I am writing primarily for those who are interested in incorporating such an exercise into their development programs.

I begin with a description of LeaderLab, so that readers will understand the context of the exercise. I then describe how the touchstone exercise works, using participants from a particular program group as examples. My discussion includes a number of observations that those involved with delivering and studying the program have made in the course of offering this exercise. Finally, I conclude with ideas for using such a technique for leadership development programs.

The Touchstone Exercise

The touchstone exercise has been a component of LeaderLab since 1990 (when the pilot programs began). During this time, I have facilitated it in approximately thirty programs with over 600 participants. Because it is important to understand the context within which it occurs, I will begin by describing the program's intent and basic structure.

The Context: An Action-oriented Leadership Development Program

The goal of the LeaderLab program is "to encourage and enable leaders to take more effective actions in their leadership situations; actions which develop themselves and others in the pursuit of a goal that benefits all." Its basic framework is this: Participants spend a week at the Center working with a set of eight leadership competencies, getting feedback, and acquiring the tools to achieve lasting behavioral change. They create an action plan, which they implement back at work over a three-month period with the help of a Center staff professional, a *process advisor*. Then they return to the Center for another week to reflect on their actions and behaviors, consolidate their learnings, and create a second action plan. The process advisor helps them maintain focus and implement their goals through a second three-month period back home. (The structure and components of LeaderLab are described in depth in Appendix A.)

In writing about the key issues encountered in designing the program, two of its creators—Bob Burnside and Victoria Guthrie (1992)—said, "A leader uses more than rational intelligence and the meanings of words" (p. 22). They pointed out that "being in a management position develops one's ability to intellectualize and verbalize, but the use of other methods of expression is neglected" (p. 24). To engage the emotions and action along with the intellect in the learning experience, a number of activities which have not been traditionally utilized in leadership development programs were adapted for this program.

These activities partly rely on nonverbal exploration in order to circumvent the logical structures people typically use to interact with each other and to construct a view of the world. For most of us, these structures are based on words and are linear in nature. It is not the aim of such nontraditional methods to throw words and rationality out the window but rather to provide participants with an opportunity to work within a nonjudgmental "laboratory" environment (hence "Leader*Lab*") to take a look at the demands of their unique leadership situation in a different way.

Three of the nontraditional program activities are acting, three-dimensional problem-solving or "body sculpting" (see Appendix A for a description of these activities; also described by Conger, 1992), and artistic activities. The artistic activities include creating a pastel drawing early in the first week participants spend in the classroom. The drawing depicts their current leadership situation in terms of family, work, friends, interests and hobbies, and self-development. They then break into small groups to describe these to each other. Participants make a second drawing when they return for the second classroom week; this depicts the successes and barriers they have encountered in carrying out their action plans during the intervening three months. They then verbally describe these to the whole group.

The major artistic activity is the touchstone exercise.

The Exercise

The touchstone exercise takes place on the last day of the first week of LeaderLab and then again at the latter part of the second week. During the first week the focus is on the individual. During the second week the focus broadens to encompass the work group. In this report the touchstone's effect on the individual leader is emphasized. Therefore, I will focus in the cases presented below on its use in the first week of LeaderLab. (A complete description of how to conduct the touchstone exercise can be found in Appendix B. Stories about touchstones constructed in the second week are contained in Appendix C.)

Setting the stage. By the end of the first week, the participants have spent five days working with the LeaderLab model's five broad leadership competencies and have spent much of their time in experiential learning activities. The day before, they met with their process advisor for an in-depth session to work through the 360-degree feedback they received, and they worked on their life biography, addressing life-stage issues, and exploring their sense of purpose in their leadership situation. This was followed by a guided-visioning exercise designed to begin the process of creating the future.

On this morning, they will write vision statements capturing the essence of their individual experience during the guided-visioning exercise of the previous evening. Their next task is to analyze their current situation and develop an action plan for the period between this session and the next, three months hence. During the time back home, participants will work to implement their action plan on the job with the support of their process advisors and change partners. When they return to the Center for their second session, they will reflect on and evaluate the actions they have taken and create new action plans. The touchstone exercise precedes action planning. This placement is deliberate and the purpose is to refocus participants on their sense of purpose in their leadership, creating a link between vision and action.

Conducting the touchstone. At this point in the program, I as the artist-facilitator introduce the exercise and offer as an example a personal touchstone of my own and its meaning. Throughout the exercise, I will provide as little or as much technical facilitation as is needed by the particular participant group.

When participants make their touchstones, which they create from a collection of varied materials, they are asked to think back on their experiences of the week and to put into their creation whatever they want most to remind themselves of once they are back in the midst of the daily chaos they will all face when they return to their jobs. They select from the pool of materials anything that speaks to them about their individual sense of purpose in their unique leadership situation.

Once they have completed the touchstone, they sit with it and record its "story" in the learning journal they received at the beginning of the week. They have already begun journaling as part of a six-month commitment to their leadership development process in LeaderLab. Each participant will take his or her touchstone home and put it in a place where it can be looked at from time-to-time during the next three months so that it can be a reminder of what the person considers most important in his or her leadership. It can be put anywhere, on a desk or in the closet, just as long as the person will look at it periodically, reflect, and recenter in his or her sense of purpose. After completing the exercise, participants reconvene in the main classroom and share their touchstone "stories" with each other. Though participants are encouraged to share the story, it is made clear that the sharing is voluntary.

The group of participants who are described experiencing the touch-stone exercise in the next section are a composite of individuals from several program groups. I felt that this was the best way that the reader could understand the sequence of events, the interpersonal dynamics, and the self-reflection that go into creating a touchstone. I will refer to them as a group (or subject group) for the purposes of this paper, though they are drawn from different programs given at different times. The group's composition is discussed below.

By design, the group is diverse: there are nearly as many women participants as men, and there is a mix of ethnicity. The twenty-one managers, executives, and administrators in the group have come from a variety of organizations: for-profit corporations, not-for-profit foundations, educational institutions, government. One is the executive director of a national information and advocacy organization for a sector of the disabled population, another is head of HR for a large government agency. One individual had recently taken over a major product division in a large, traditional manufacturing company. Another is chief counsel for a regional power utility. One runs a university graduate management program, and still another heads up R&D for an innovative high-technology company.

I start with a description of the subject group familiarizing themselves with the materials and process of the exercise. For the purposes of providing a personal view of the entire process, two actual cases (Maury and Jessica [the names are fictional]) are recounted.

Participants' experiences. On the day of the exercise, the participants in our subject group gather at a large table in the center of an exercise room, which is heaped with a profusion of mostly natural, some man-made materials: twigs, bark, dried leaves, acorns, sweetgum pods, polished stones, shells, shiny papers, moss, ribbons, tree ears, small stumps, dried vines, rocks, driftwood, colored glass, sheet copper. On paper-covered worktables against the wall are glue guns, glass cutters, wire, tape, string, pliers, wire snips, scissors. Participants begin to pick things from the pile—digging, reaching, discarding, exchanging, poking around. A few people stand back, hovering, half-bashful, half-expectant, until the first special thing catches their eye.

"Pick anything you want," I tell them, "and as much of it as you want. Pick anything that speaks to you about what's important to you in your leadership. As you start to put your touchstone together, be aware of the meaning of each thing and about the relationships of the things to the whole."

"I need something sharp," someone says. Someone else asks, "Is there another pinecone?" "This is perfect!" "Is it okay if I cut this?"

One or two are still waiting for inspiration. Others are already at worktables, sitting, standing, hunching over small sculptures taking shape out of disparate objects carefully selected and painstakingly assembled. The participants and I go back and forth, discussing the details of accomplishing the task:

"How can I stick these together?"

"That caulk might work if you tape it in place for a while."

"This twig keeps falling over when I glue it to the base."

Participants examine materials

"The surface area might be too small for the glue to hold it upright. Maybe it needs something else to support it."

"I couldn't find anything to represent my . . ."

"Keep looking: think about the qualities it has and try to find something with those qualities. You'll see it."

"Will these stick together?"

"Look for something like a piece of wood to fill in the void between that curved piece of bark and the rock—you'll get a better contact with the glue—and watch the glue gun, it's hot!"

The stragglers scan tentatively over the table, finally selecting one object, then another. In the midst of all the activity, two individuals come into focus. I have chosen them to discuss as cases because each approaches the process differently, but each learns a powerful lesson. We join them as they work to conceptualize and design their touchstones.

Case No. 1. Maury is head of a business unit within a global finance institution, and at thirty-one, is the youngest person in the room. He is on a fast-track in his career, a high-performer whose major problem area is inter-personal dynamics. Although aware of this from feedback he has received, he has never succeeded in his attempts to change his behavior.

Maury came in to the exercise room last. He has been having difficulty with the less traditionally content-focused segments of the program, hanging back during the two "acting-leader" segments earlier in the week and making mildly sarcastic comments during "three-dimensional problem solving." He participated, but without enthusiasm, until his feedback session with his process advisor. After that, something shook Maury up so much that he looked like a different person. Instructors and other participants asked him if he was all right, then gave him a respectfully wide berth when he said he was "working on something."

Maury stands at the table for a long time, looking thoughtful. "Pick up anything that speaks to you about your sense of purpose in your leadership," I tell him. He chooses a piece of plywood which has an irregular rectangle as a base for his touchstone. The rectangle is right-angled at one end, but the corners have been cut off the other end. After some more thought he picks up two clam shells. At the worktable, Maury glues the two clam shells side-by-side at the right-angled end of the base. One is a medium-sized half of a shell, placed inner side up. "Open," he says, "it represents my analytic side." The other is a small, whole clam shell, only slightly open. He uses a small stick to pry it open just a little further. "This is my emotional side." He returns to the table and selects a dried leaf, which he takes back to the worktable and places

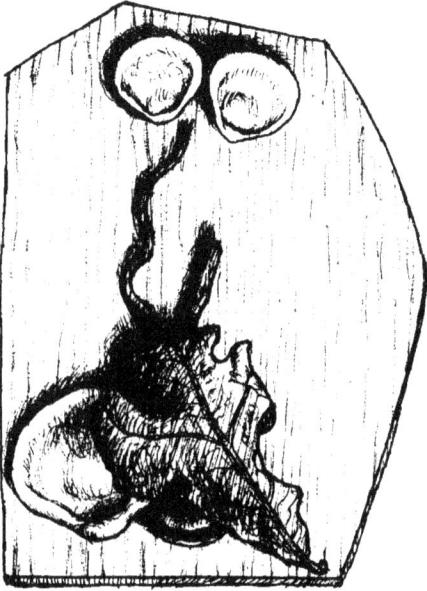

Maury's touchstone

so that it almost completely covers the smaller shell. "My emotional side is covered up," he says. He returns again to the table and looks intently until he finds two more clam shells, of nearly equal size. These he places side-by-side at the irregular end of the base, inner side up, saying, "This is my analytic side and my emotional side, more in balance. This is where I'm coming from, all right angles and analytic, and this is where I want to go—away from all the right angles." He looks at me and adds, "But there's something missing—how do I get there?" I say, "Is there a path from here to there? If so, what might it look like?"

Maury goes to the table, and quickly this time, selects two more objects: a twisted piece of dried woodvine, which he breaks to a selected length, and a smooth, straight, finger-sized twig with the bark removed. He places the smooth twig so that it touches the side-by-side shells at the right-angled end of the base and points at the shells at the opposite end. Its shortness does not permit it to touch the shells at the other end. The twisted vine he places next to the twig on the base, and the other end reaches to the shells at the other end of the base. I ask, "Does it need anything else, or are you finished now?" He responds, "It's all I can deal with emotionally right now."

Some people finish their touchstones very quickly and return to their seats after twenty minutes or so. As instructed, each sits with his or her touchstone for a few minutes to "let it tell you its story," then write that story in their personal learning journals. Maury and a few others take forty-five minutes to finish their touchstone—every minute allowed for the exercise— before returning to their seats. As the group formally reconvenes to share touchstone stories, everyone gets up for a "tour" of the touchstone "gallery," to examine everyone else's work. They're asked to be silent during the tour, but rarely are. They tease each other and several excitedly share their stories with others. It is hard to get them all back into their seats, but when most finally are seated, I ask if anyone is willing to share his or her touchstone story with the group.

After two people have told their stories, Maury shares his. "This piece of plywood is me. See, it's right-angled at one end—that's where I'm coming from, very analytic and hard-edged. This big, open shell is my analytic side. No problem with that, it's well-developed and I rely on it a lot. But here, next to it is my emotional side. It's a lot smaller, only open a little way, and it's pretty well covered up. The other end of the base is not right-angled, it's irregular. These other two shells here at this end are more equal in size. They represent my analytic side and my emotional side, more in balance. That's where I want to go. But how am I going to get from here to there? There's a straight path here at the right-angled end. It's fast and smooth, but if you notice, it doesn't get me all the way to the other end. This other path, the twisty, turning one, does go all the way. It'll take me longer. It's winding and rough, but it will get me there."

That day, Maury and I sit together at the lunch table. "You know," he says, "when the staff introduced themselves the first day and here was this 'artist' who was going to be working with us—I sat there and thought, 'Geez, I need *this* garbage?' Now I think it might have been the most powerful part of the whole program for me." He later writes to say he has added to his touchstone, placing a series of pebbles along the path symbolizing the obstacles he recognizes he will meet along the path of his ongoing development.

Case No. 2. Jessica is a British manager in a multinational pharmaceutical company. She has been feeling the stress and frustration that followed a transatlantic merger between her former British company and an American

Jessica's touchstone

one. Jessica creates her touchstone without asking for assistance. Her demeanor during the exercise is one that seems to reflect some ambivalence toward the potential value of the exercise. She works quietly, without any outward show of enthusiasm. When I tell her that the small polished amethysts she has placed on her touchstone are symbolic of sobriety and spiritual growth, she responds in an ironic tone, "Perhaps I chose the wrong stones."

Like Maury, Jessica shares her story with the participant group. Its elongated hexagonal plywood base is "shaped like a chemical structure." She points to a piece of red glass set on its side, forming a wall or barrier between two groups of small stones. The red glass barrier is the single largest object on her touchstone.

"Today we are divided by the Atlantic and the history of the two different companies before we merged," she says. "There are individuals on each side"—she points to the different colored stones, including the amethysts, divided by the glass barrier and flanked by large, spiky thorn twigs—"and there is a prickly atmosphere." She points to bark fibers, representing earth, and a dried rose blossom. "If we—I—can sow the right seeds, we could get a rosy future where the company would have many jewels [a small heap of semiprecious polished stones]—products, people, processes—and at the same time there would be time for sea [a piece of blue glass glued flat to the base at one end] and sunshine—we can all go to the seashore!" she laughs.

In the second week of the program (three months later), when participants return to assess their progress on goals they had set for themselves in week one, Jessica reports, "I realized that some of the frustration I was feeling was because I wasn't doing enough listening. In the last three months I've made an effort to talk individually to my peers, and as a result I got a different picture of my own potential for effectiveness. Now I see myself as active across the board." When she makes a second touchstone, the opposite sides of the Atlantic are still represented, but Jessica depicts an active exchange across the medium that she previously represented as a barrier.

For more cases, see Appendix C.

Discussion

What does the touchstone exercise accomplish for the managers and executives who go through LeaderLab? What impact does it have on their effectiveness? Since its inception in 1990, over 600 participants have attended the program. During that time, researchers, program facilitators, and I have had a chance to measure, observe, and reflect on what outcomes the exercise has produced in participants and how it compares to other program components. Six of these observations are discussed below: the effectiveness of the exercise, how the exercise has changed over time, the similar themes that have emerged in the touchstones, the lessons learned that have been used back home on the job, the use of the touchstone experience to stimulate a story, and resistance to the use of the artistic component.

Effectiveness of the Touchstone Exercise: The LeaderLab Impact Study

We were interested in knowing what kind of impact artistic activities (drawing and the touchstone exercise) had on program outcomes. A Center researcher, Dianne Young, observed and assessed the impact of all of the action-based components, including artistic activities. Her study aimed to answer three key questions about the effectiveness of an action-based leadership development program: (1) What are the action-based design features of such a program? (2) What are the outcomes? and (3) How do participants integrate action, learning, and change? Both qualitative and quantitative data about action-oriented features were collected three-to-four months after the program from thirty-two participants who attended sessions between March 1992 and February 1993 (Young & Hefferan, 1994).

The study found that the program is having positive impact (Young & Dixon, 1995), and that artistic activities contribute to that impact. The latter finding was illustrated in the participants' rating of the effectiveness of the action-based features. Three months after the program, participants were asked to rate each of these features on a scale of 1 to 10 in terms of its helpfulness in reaching their goals (Young & Hefferan, 1994). Components rated in order of effectiveness were: action planning, process advisor, program structure, artistic activities, nature walk, acting, journaling, visioning, change partners, and group sculpting (see Appendix A for descriptions of these activities). Artistic activities were rated 7.1; this compares to an average rating of 8.9 for the process advisor, rated the single most effective program component, and an average rating of 5.7 for group sculpting (three-dimensional problem solving), the component rated least effective.

Young compared these data with evaluative data collected during program sessions. She found that artistic activities rated relatively high in the quantitative measures, but she also discovered that, along with the process-advisor role, ratings for artistic activities actually rose three-to-four months after the program (Young & Dixon, 1995). This is illustrated in Table 1, which shows the four top-rated components and their ratings during the program and then three months later.

The evaluation outcomes and the integration of action, learning, and change are important in understanding the context of the finding that artistic activities add to positive impact. Using in-depth interviews and a "change questionnaire," based on the interview data and developed specifically for this study, positive outcomes were reported in terms of observable behaviors and actions taken in a number of areas, notably trade-offs (the capacity to make decisions in situations which are not clear-cut), self-assessment and seeking feedback, interpersonal relations, facilitating communication, organizational systems (the capacity to see the "big picture" and its interconnections both within and surrounding the organization), and flexibility (Young, 1995).

Table 1
Ratings of the Top Four Action-based Program Components

Program Component	Average rating across 6 months of LeaderLab program*	Average rating 3 months after completion of LeaderLab**
Action planning	9.0	7.8
Process advisor	8.6	8.9
Program structure	8.6	8.2
Artistic activities	6.6	7.1

* 7 public runs of LeaderLab (*N* range = 52-118)
** 3 public runs of LeaderLab (*N* = 32)

As to how participants integrate action, learning, and change, the researchers found evidence that participants approached implementation of action planning from three different models: goal focus, vision focus, and process focus. The goal-focus model takes action planning "as a blueprint for specific actions" and "is reflective of the way action plans are most commonly referenced in the literature: goals to be attained." "In the second model, vision focus, the respondents seemed to view the items on the action plan as beginning steps for a more extensive plan," taking advantage or responding to opportunities in the environment, so that actions changed and "the vision seemed to evolve as well." Young and Dixon (1995) reported that this model seemed to be the most prevalent among participants in the study.

A third model was more process-focused: "Participants viewed action planning as an ongoing process that involves acting, reflecting, and revising actions"; "initial steps written on paper were viewed as just an iteration of a process that would continually evolve." These differences in focus, according to the researchers, "challenge the way leadership and management-development professionals have typically viewed and presented action planning to their participants. It has most often been framed from the goal-attainment model, assuming that an action plan functions like a road map to get the participant from point A to point B.

"These data suggest that for many participants action plans are much more dynamic" (Young & Dixon, 1995). This variation in approaches to action planning is likely influenced by, among other factors, the process perspective of the program, in which action plans are constructed, then implemented for three months on the job, then revised and implemented again, emphasizing a dynamic view of action planning.

Although the researchers did not systematically study the relationship between the touchstone and action planning, there is evidence that participants found the touchstone helpful in working toward their vision. Comments from participants show a perceived link between the visioning component of the program and their touchstone representation of that vision. This link often extends to action planning. "It reinforced the power of pictures and links in with the visioning. It gives more power to the message." "My personal plan was to stop being such a lawyer. It [the touchstone] helped me be more emotional and more open, less analytical." "The touchstone was great. It helped me clarify my vision in ways that thinking, writing, and verbalizing couldn't do." "The touchstone will be the most useful for me because it involves emotions as well as thought and will be a lasting, practical reminder of how to move forward and a check on where I'm going."

According to Young and Dixon (1995), "The way the program is able to deeply personalize the action planning and artistic activities may be the key to their impact. Finding a way to connect to some deep personal meaning through the artwork and then carry that into the action planning may increase commitment and help integrate the participants' own vision with their leadership situation at work."

How the Exercise Has Changed Over Time

Although the touchstone exercise itself has not changed since the first program—participants still create an object representing what is most important to them in their leadership—the placement and emphasis of the exercise has changed somewhat over time as the program team has come to better understand its effects and value. All three of the major nontraditional components mentioned above (artistic, acting, and group sculpting) were incorporated into the program on an experimental basis and have undergone some changes, although none has been dropped or replaced. Most participants think that the three add value to their leadership development process.

In the first year of the program, the touchstone exercise was run only once during the classroom sessions, at the end of the first classroom week. Subsequently, a second touchstone exercise was added, to the end of the second classroom week. Participants create a touchstone representing the shared purpose of their work group. To help participants integrate input from members of their work groups, who are not actually present in the classroom, a survey was added. It asks respondents in the back-home work group to create metaphors representing the shared purpose of their work group and the leadership qualities of the participant, which they feel contribute to this shared purpose. Surveys are distributed by the participant and returned to the LeaderLab program coordinator, to be returned to the participant during an exercise in the second week. Their use is optional, but most participants opt to use them, and the response rate has been relatively high. The metaphors help the participant to create a touchstone which then integrates the contribution of stakeholders in the work group's shared vision.

More recently, as discussed above, findings of the LeaderLab impact study strongly suggest a link between the concrete, personal metaphor or vision of change represented in the touchstone and positive program impact over time. The emergence of these indications from the study have resulted in a change in the flow of activities on the program's last day in both classroom sessions. Participants write a "first cut" of their vision statement, then create a touchstone representing both their vision and what they want most to take back to their leadership situation from the program experience. During the introduction to the touchstone exercise, the links between vision and action plans are more clearly articulated. Participants then revisit their written vision statement, integrating anything which might have emerged as insights from the touchstone exercise. Then they set their action plans, followed by a small group consultation on action planning with their in-class change partners. Many participants bring their touchstone with them to this consultation. This progression is meant to emphasize the link between vision and action. Evaluative comments received since this change suggest that it has increased the practical value of the exercise.

Through future research, we hope to learn much more about how artmaking relates to how participants integrate learning, change, and action.

Common Themes in Touchstone Representations

Over time, certain common themes and imagery have appeared in the touchstones. One is a representation of self, frequently in terms of a hoped-for change, for example, "This rock is what I was. It's hard and impenetrable. This pine twig is what I'm becoming: more alive and flexible, willing to bend when necessary."

Another common theme is the path from one state to another, representing the process of change, complete with prickly obstacles and the difficulties the participant expects to encounter in the change process. Something unseen or hidden is often represented, "something I can't see now but I know I'll be able to see later on," or "something that's affecting my [behavior or situation] that I haven't been aware of," or "something that is meaningful to me but I don't let it show very often." Listening, being open to feedback and ideas as opposed to being closed, "planting seeds" in the work environment, and growth, both personal and organizational, are all commonly depicted themes. Participants often incorporate objects celebrating the diversity and affirming the importance of the unique contributions of people in the workplace.

Many of these themes and images reflect the behavioral feedback participants have received in LeaderLab and represent metaphors for personal-development goals based on that feedback, as well as a commitment to these goals. The dot, symbolizing sense of purpose in the LeaderLab model (see Burnside & Guthrie, 1992), and the spiral, representing learning and growth, are sometimes represented in participants' touchstones.

Because of the three-dimensionality of the touchstone, relationships are vividly depicted not only in the meanings of each object in the representation but the spatial relationships between the objects and the whole presentation. Some participants have created touchstones which show a different "picture," depending on the angle from which they are viewed. Relationships are depicted between self and others, between what is seen and what is not seen, between "here" and "where I/we want to go," using all the dimensions of above, below, behind, in front of, surrounding. Some are linear in depiction (such as the path from here to there); many are multidirectional and multidimensional. All of this becomes part of the "story" of each participant's touchstone and is reflected in the language he or she uses to describe it.

Touchstone Lessons Used on the Job

Very often, the effect of the touchstone is reported only after participants have returned back to the work environment. In the second week of the program (conducted after a three-month interval), participants return to report on their progress in working toward the goals they set for themselves in the first week. During the first week, Frank, a railroad executive, presented himself with a gruff and sometimes intimidating manner. He did not build a touchstone (which is unusual but not surprising, considering the difficulty Frank had all week with "the soft stuff") but instead picked up two shells from the table. I asked him what they signified, and he said they would remind him to listen and "not beat up on people." He shared the same terse story with the group. His in-class change partners, fellow participants with whom he had discussed the feedback results he received from his work group back home, knew that this was his major challenge and goal. When Frank returned for the second week of the program, he greeted me and others by opening up his briefcase and displaying the two shells, still inside. His report to the group was that he has succeeded in improving his listening skills, his interpersonal relationships, and his effectiveness generally.

Paul, senior vice president and chief counsel of a mid-Atlantic electric utility company, depicted himself in his touchstone as a pinecone. "This is me," he said, "rigid and prickly." He also included a young cypress twig. "This is what I want to be: soft, flexible, a branch from the 'tree of life.'" When he returned for the second week, Paul said to the group, "I'm less defensive, less judgmental, all the things you'd expect a good lawyer to be!" He laughed. Paul continued: "A lawyer friend told me, 'I'd be scared to do what you're doing—you've been successful in corporations being the way you were.' But I came here in the first place because I knew I needed to change. I know now it all depends on interpersonal relationships." (Paul also kept a promise to himself to return to his abandoned hobby of painting in watercolor—as proof, he sent a small painting of his touchstone.)

Maria, also a lawyer, in a Fortune 50, global, high-technology company, made an all-blue touchstone. "My feedback was that I'm aggressive. So I made a blue touchstone. I keep it on my desk. When tempers are going to flare, or I'm being competitive or feeling insecure, I glance over and it reminds me to calm down. I would normally be 'red.' Being 'blue' has helped

Robert's touchstone

me in my relationship with our new general counsel. It also helped me with my two bosses, who have different styles and often put me in the middle."

During the program, Robert, marketing manager for a Native-American-owned electronics and communications defense subcontractor, created a touchstone which dealt with being between two worlds, "the Indian and the non-Indian." He debated whether to place a bridge between the two worlds but decided not to. "It doesn't need a bridge if I'm the base." Though he depicted the two worlds as separated by a barrier of cultural difference, "We both look at the same sky," he said. This is an obvious idea, and I asked him if he never thought of it before. "Yes, I have," he answered, "but I never *saw* it before." Robert reported that he has been able to internalize this idea of harmony in a way that can be expected to aid him in his daily decision-making and interpersonal relations back at work.

Melinda used her touchstone "to focus my own vision—then I had my subordinates create a sort of 'group touchstone' when we did our corporate-visioning project. Everyone brought an object that to them represented the values that were most important. Everybody feels ownership in the end result."

Although few have reported making use of the technique itself like Melinda has, participants very often report using their touchstone to communicate their vision to others in the organization. Some keep it on their desk and respond to the inevitable questions ("What on earth is that?"), using the opportunity to share their program learnings and vision. Several use the touchstone right away, for example in meetings called to communicate newly clarified vision and goals to organizational constituents.

Tom is an executive of a financial company which has been going through traumatic downsizing and bad press. His touchstone is a large, curved piece of bark mounted at a graceful angle resting on another piece of wood. He had glued several more small pieces of bark onto the wood and the base. "This is our canoe. You can see it has run aground, and it's kind of broken up. But look down here at the base and

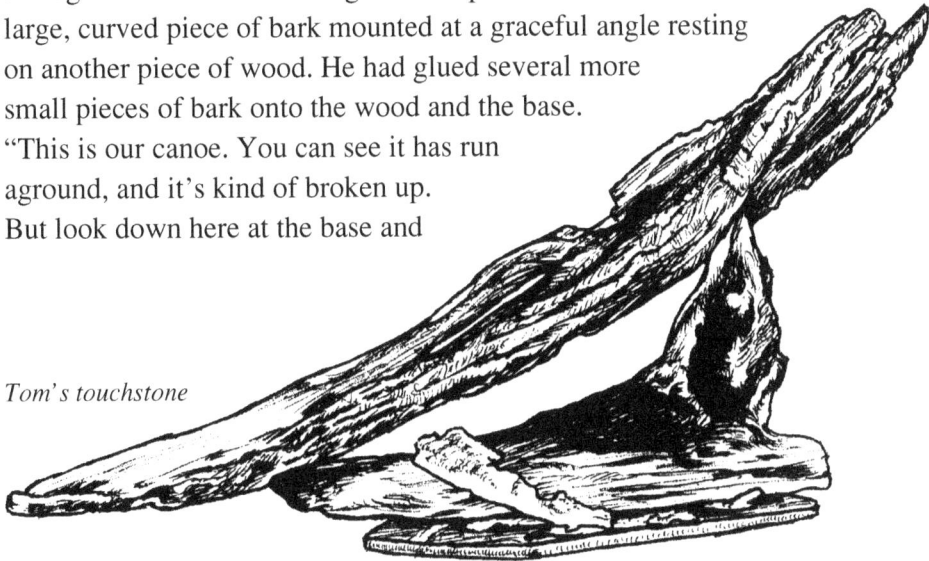

Tom's touchstone

you can also see a lot of small pieces of bark that have broken off. They represent the abilities and commitment everybody in our team brings to the table. I know that there are enough of them to repair our canoe. Together, we'll be able to get it going again. It won't be as pretty as it was, but it'll get us there!"

Guillermo, a European manager in a global manufacturing corporation, had just made a transition to head a Latin American division. He chose a pinecone to represent a concept of team and community growth and interdependence. The pinecone symbol is a familiar and well-understood metaphor in his home culture. He intends to import it and its meaning to his new location to symbolize his vision for the cultural change he has been mandated to create in his new position.

The Touchstone Exercise as a Story

Although the touchstone begins as a nonverbal exercise, the story that comes out of it is at least as important as the object itself. Often, groups express surprise at hearing the qualitatively different nature of the language which emerges in the storytelling. The language is poetic and metaphorical, often expressive of a spatial sense of relationship which emerges directly from the three-dimensional relationships in the touchstone and of the simple qualities of objects used in it (it can be difficult to talk in an abstract, analytic way about a pinecone).

LeaderLab is designed to maximize the learning participants receive from each other. The touchstone storytelling becomes an opportunity for learning from a qualitatively, often startlingly, different kind of discourse from the usage that is the daily norm for most of us. Probably the easiest way to express the difference is that it sounds human. Participants often share more personally than they have experienced from one another previously in the program.

In a contract program conducted for an all-male, high-potential group in a multinational manufacturing company, the public telling of the touchstone stories assumed great importance. Each member of the group wanted to hear every other member's story. They felt the touchstone stories to be personal and more revealing of their feelings about what really, deeply motivated them in their leadership than anything else they had experienced during the three years they had been working together developmentally.

Many trainers and increasing numbers of theorists are making use of vivid metaphors to convey important concepts about the changing nature of organizations and leadership roles. An example is Peter Vaill's metaphor of "permanent white water" for the organizational realities of the modern world, and the approach to the kind of leadership demanded in this white water as a "performing art" (Vaill, 1989, p. 2).

The touchstone exercise begins with the metaphor of the artist's practice and ends with the creation, through active engagement in artistic practice and process in the laboratory setting, of new and personally meaningful metaphors that are unique to each participant's situation. A participant in a recent program worked with his process advisor to create a metaphor to guide him in changing adverse behaviors he had been alerted to through his feedback. To express his goal of remaining calm and staying in his chair long enough to listen to subordinates, rather than leaping into action and trampling all opposition, he pictured the seated figure in the Lincoln Memorial. He used a penny in his touchstone to remind him of this image of calm, listening, accepting repose. When he returned to the second classroom session, he reported that his touchstone, which he kept on his desk, was very important to him in focusing his change efforts and that this metaphor of calm would continue to guide him.

Resistance to Artistic Activities

Initial discomfort and resistance such as Maury's to using nonverbal techniques is not at all uncommon and can be expected among many of the participants on first exposure to them. We refer to this discomfort as a "GAG": going against the grain of accustomed behaviors (Bunker & Webb, 1992).

We think that the touchstone exercise can be helpful in the process of personalizing learning and giving concrete impetus to action planning. However, it may not work for everyone, for different reasons. Although most people not only participate in the exercise but report that they get at least some value out of it, to date two LeaderLab participants have declined to create a touchstone.

In one case the participant felt it hypocritical to represent positive aspects of his leadership vision because he was on that day dealing with personal feelings of distrust for the motives of his organizational superiors. I suggested that he find something to represent this issue. Like Frank, described above, who selected shells but did not build a touchstone, this man accepted the suggestion and chose a piece of iron pyrite, "fool's gold." When the group shared their stories, this participant did too. Holding up the iron pyrite he spoke about his commitment to work within the organization and support its broad goals as long as he saw the value in them but not to be distracted from his personal sense of values by the fool's gold of overcompensation and values spoken but not acted on.

In a second case a participant who had created a touchstone in the first week of the program, representing his individual sense of purpose in his leadership, declined to build a touchstone in the second week which would represent the shared purpose of his work group. When participants shared their stories, I asked him if he would be willing to tell the group his reason. He said, "I don't need an object to crystallize my thoughts. My work group is dynamic, not inert. I would make a group touchstone [which all members would participate in creating]." In an artistic exercise earlier in the week, he had made a drawing that captured to his satisfaction an image of the successes of his group over the previous three months.

Another participant who was felt by trainers and process advisor to be emotionally disengaged during the week, spent the full exercise period creating a touchstone of unusual density and artistic quality (indications of artistic quality in this example involved the use of a large amount of varied materials to create a composition with a sophisticated and harmonious use of spatial relationships, color, and textures). When invited to tell his story, he said, "It doesn't mean anything!" He did not care whether it was sent home for him or not, and it appeared clear that he would not make further use of it if it were.

There have been some participants who, although not expressing initial resistance, simply did not get much out of the process. A sample of written evaluative comments includes the following: "Least useful: touchstone. For me, it seems that you can read anything into anything. However, having said that, it was good to reflect on what is important while gluing together a collection of stuff." "The touchstone was fun at the time, but didn't have a lasting effect. I think they probably serve value but didn't work as well for me." "I'm not sure the body sculpting and touchstone worked for me, although I think they helped others and it was enjoyable." The conciliatory, almost apologetic note in the comments, which is almost universal in negative comments returned on this segment, may be a consequence of the personal nature of the presentation by the trainer and a resulting reluctance on the part of the participant to offend. At the very least, most participants, after some mild initial anxiety ("I was nervous at first, but then it just seemed to come together") do seem to enjoy the process.

Other participants have felt that their own shortcomings have prevented them from getting the most out of the exercise. In a letter I received from a participant following a program, he made the following comment about his reluctance to share his touchstone story publicly. "I guess I felt like the little boy in fourth grade who was laughed at by his friends and teachers. I didn't want to share what I could be ashamed of." Another participant told me after the touchstone exercise that he had felt reluctant to share his story because he did not feel his touchstone was artistically "up to the level of production I was seeing" in the other participants' touchstones, and this embarrassed him. From many casual comments by participants, it is painfully apparent to me that many people suffer from feelings of inadequacy in approaching a mode of creation they may not have practiced or felt comfortable with since childhood. An art therapist and educator, Peter London (1989) points out that because of impoverishment of our education in visual expression, "there are some disabling myths about what art is, how to do it, what is good art, and what art is for, that have gagged generations, depriving them of a significant and natural means of expression."

The Use of Nontraditional Classroom Components in Development Programs

Can nontraditional techniques enhance leadership development programs? I believe they can. I will begin by making some general statements about their use in such programs and then discuss more specifically the roles that artists, emotions, and program configuration play in the delivery of nontraditional program components. Finally, I will conclude with some thoughts about the transferability of the touchstone exercise to other settings and on the relationship in general between art and leadership.

Artmaking, acting, and body movement methods have long been used in diagnostic settings and in family therapy. More recently, these methods have been applied in organizations seeking creativity and innovation. LeaderLab is among the first to use such methods in leadership development. The most startling things reported by people who use visual and other non-verbal methods of ideation and communication is that every one of us has the ability to do it, easily, once initial resistance is overcome. Joan, a no-nonsense entrepreneur and head of her own career-consulting business, said, "You gave me permission to do something it wouldn't have occurred to us we could do. I went into this exercise with a sense of dread, but you said, 'Don't think about it, just do it.' It was liberating and I really enjoyed myself."

As Peter Vaill observes (1989, p. 125), "The arts stand open to the possibility that the individual person is capable of spontaneously generating astonishing new material, material that goes far beyond what anyone imagined was possible. It might not be an exaggeration to say that the expectation of such periodic breakthroughs is the most powerful thing about the arts and what gives them their most profound human meaning. They demonstrate thrilling things about human nature. In this age of permanent white water, where the basic viability and effectiveness of organizations are everywhere in doubt, we are in acute need of similar breakthroughs in all sectors of society."

The Role of the Artist

From the first offering of the LeaderLab program, the role of the artist was differentiated from the rest of the trainer roles. We felt that the artist role added legitimacy to the holistic approach of the program, by having a working artist available to facilitate artmaking exercises. The presence of a working artist and a working actor as trainers supports the program's aim of helping the participant look at his or her leadership situation from different perspectives.

The touchstone is a personal expression, and the artist's work is centered in drawing on a personal "take" on the world. Likewise, I believe that the most effective leader brings all of him- or herself to work. Therefore, when I first introduce the exercise I share a "personal touchstone," an artwork of my own, and tell its "story" to the group. The work and story are chosen to model a level of personal disclosure, and its content concerns how this artwork reflects a self-development process of my own.

Although the element of diversity and professional experience is important in the artist's role as trainer, the trainer's ability to speak the language and understand the issues facing people in organizations is crucial. Both artists who have trained in LeaderLab (another artist conducts the touchstone exercise at the Center's San Diego branch) are also experienced in organizational training and group facilitation. This is a critical, though unusual, combination of talents.

The exercise has been successfully conducted by non-artists and should not strictly require the special talents of an artist to be effective. An additional approach might be for an experienced trainer/facilitator to conduct the exercise in tandem with an artist to gain the benefits of both if an artist is not available who is also an experienced facilitator.

The Role of Emotions

One reason that facilitation experience can be critical to the effective use of nontraditional techniques is that opening themselves up to new experiences is a key factor in how useful these components can be for participants. In the words of one, "[The acting session] forced me into experience I would avoid. [You] have to try it to get a benchmark—push on the margin of what you're comfortable with—a sense of what it's like to change." This opening up can also often elicit an element of emotionality in participants. After one touchstone exercise, a participant told me, "I'm sorry I didn't share my story with the group, but I was afraid I'd burst into tears." One individual mentioned above who had difficulty with the exercise because of embarrassment and feelings of inadequacy, privately gave me some direct verbal feedback following the touchstone exercise. He said that he had felt pressured by me to tell his story (he was the only member of the group who had not told a story) and that I had violated the statement I had made at the beginning of the exercise that no one need share their story publicly.

A second comment to the same effect was returned by a second participant in the written evaluation of a different program group: "The process was useful until [I was] singled out as not having participated. The process based upon initial instructions was to make it for self, share if one so desired. Maybe having gotten too into it, I felt a certain intrusion." The personal nature of the exercise can result in some discomfort. The reluctance to share, for whatever reason, must be honored.

In their program feedback report, Young and Hefferan (1994) report on participant response to the session on group sculpting that "several participants talked about the power/shock of the experience. In fact, this was mentioned most often in the participant responses. LeaderLab staff should be aware of the risk involved in these exercises." Careful introduction, linkage to program content, and sensitivity to participant resistance, doubts, and questions is required to create a positive and safe atmosphere for the session. The familiarity of all program trainers with these sessions and their availability during the week to answer participants' questions are also helpful. This said, the fact remains that the power of the nontraditional components is in experiencing them, and this cannot be transmitted in words.

A specific emotional response of many participants to all three of the nontraditional components mentioned above is "fun." This should not be undervalued in terms of its usefulness for people in leadership positions in organizations: Possibly the most common leadership challenge mentioned by participants in LeaderLab, next to the need for facing change, is the need for personal balance.

The Role that Program Configuration Plays in the Effectiveness of Nontraditional Components

The findings of the impact study (Young & Dixon, 1995) suggest that the embeddedness of artistic activities in the program is a key element of their effectiveness. In keeping with the central action-based, "laboratory" concept of LeaderLab, the artistic and other nontraditional components have been used to enhance traditional methods. All of the program modules and components are placed within the program flow to complement and reinforce each other. The use of nontraditional components is introduced early (on the first day) and they generally alternate with more traditional modules (such as short lectures and small-group work), giving participants an opportunity to experiment with new behaviors and process learnings. The incidence of sessions incorporating nontraditional activities throughout the week may also help to reduce the "GAG factor" involved for some participants.

Transferability

Although the touchstone exercise was developed for a specific purpose and is embedded in the overall design of LeaderLab, the exercise and variations of it have been used in other settings, by myself and by trainers who are non-artists. It is typically used near the end of a program or workshop, to embody commitment (to such things as action plans, self-development, and personal goals), and it can also be useful in group development by providing a safe forum for personal sharing. The nature of this sharing is altered by the initial nonverbal experience of the exercise—it provides an avenue to deeper feelings and connections, giving participants a different picture of each other than might otherwise be possible in a short time. The exercise has been used in public-program contexts, involving participants from different organizations, and in contract programs, involving individuals from the same organization (examples: financial, manufacturing, public utility, construction). It has also been used with school children.

The exercise can be individually focused or focused toward group goals. Single "group touchstones" have been collaboratively built to represent an expression, and experience, of group cohesion and commonality. The feel of the exercise in this context is celebratory.

The uses for nontraditional artmaking activities in addition to leadership development are: stimulating creativity, developing shared vision, strategic goal setting, and team-building. (See Appendix D for examples.)

Art and Leadership

Many writers on leadership have pointed out a connection between art and leadership. According to Max De Pree (1989, p. 3), Herman Miller CEO emeritus, "Leadership is an art, something to be learned over time, not simply by reading books. Leadership is more tribal than scientific, more a weaving of relationships than an amassing of information, and, in that sense, I don't know how to pin it down in every detail." In these remarks, he encapsulates four aspects of common wisdom concerning art: that it has the quality of being irreducible by analysis to a mere sum of constituent parts; that it has to be experienced directly; that it continues to have something to say to us over time; and that it is in essence about perceptions of connectedness, interrelationship, and integration. Recent thinking on leadership as "meaning-making in a community of practice" (Drath & Palus, 1994) potentially incorporates all of these aspects as part of meaning-making. Leadership, like art, then can be seen as a process instead of a set of discrete practices.

In support of thinking of leadership as process rather than procedure, both Abraham Zaleznik and Peter Vaill warn against conceptualizing a leader's purpose as a managerial "list of functions." In an age of what Vaill (1989, p. 2) calls "permanent white water," leaders are required who will engage the talents, imagination, and commitment of their constituents. According to Vaill, who conceptualizes leadership for the future as a "performing art," a leader's action-taking is "a concrete process performed by a whole person in relation to a whole environment populated by other whole persons (that is, not other lists of functions). This whole process is embedded in time and is subject in the real time of its operation to all the turbulence and change that surround it, that indeed suffuse it, because the turbulence and change are within action takers as much as they surround them" (p. 114).

Zaleznik (1992) believes that "business leaders have much more in common with artists, scientists, and other creative thinkers than they do with managers," and that a central feature of that commonality is that "leaders tolerate chaos and lack of structure and are thus prepared to keep answers in suspense, avoiding premature closure on important issues.

The touchstone exercise, because it is rooted in the common ground between art and leadership, can help leaders respond to turbulence with creativity. As Rollo May suggests, "What if imagination and art are not frosting at all, but the fountainhead of human experience?" (quoted in London, 1989).

This question opens new avenues for exploration in both artmaking and leadership development studies and practices. The relative ease with which hundreds of participants in a leadership development "laboratory" have created art objects with deeply and unexpectedly meaningful stories suggests that, like a fountainhead, artistic expression is close to the source of human experience and can be regarded as an innate capacity which has been underdeveloped and underutilized by most people in our society. The relevance of that capacity to effective leadership is an open question. However, in a rapidly changing and chaotic world, leaders must develop capacities which enable them to exceed the natural limitations of linear, analytic thinking and conceptualizing.

Artmaking is a process of exploration and inquiry that connects us with what is most deeply held within each of us. It connects us to ourselves, our environment, and to each other in ways that can reveal fresh insights and perspectives as well as inner touchstones that hold value over time and remain unshaken by chaotic events. Artmaking and leading are, above all, profoundly human activities.

Note

Maury's program experience is described in a case study in Jay Conger's *Learning to Lead* (1992, pp. 181-186). Conger interviewed Maury together with his process advisor following the program. In the session with his process advisor in the program's first week, Maury indeed experienced dramatically emotional personal insights. In a journal entry following the meeting, Maury wrote: "Using your whole self—complete with the emotional side—appears to be what I need. Why am I in a hurry?" (Conger, 1992, p. 184). Following the program, Conger notes, "The danger, then, was that he would give in to feeling pressured again and revert to old behaviors. But he resisted this in several ways. In meetings at work, he was deliberately modest and low-key, which was very different for him. He delegated more. He asked people their opinions. When you're known for being fast and having the answers, that is very difficult to do" (p. 186).

References

Bunker, K. A., & Webb, A. D. (1992). *Learning how to learn from experience: Impact of stress and coping* (Report No. 154). Greensboro, NC: Center for Creative Leadership.

Burnside, R., & Guthrie, V. A. (1992). *Training for action: A new approach to executive development* (Report No. 153). Greensboro, NC: Center for Creative Leadership.

Conger, J. (1992). *Learning to lead: The art of transforming managers into leaders.* San Francisco: Jossey-Bass.

De Pree, M. (1989). *Leadership is an art.* New York: Doubleday.

Drath, W. H., & Palus, C. J. (1994). *Making common sense: Leadership as meaning-making in a community of practice* (Report No. 156). Greensboro, NC: Center for Creative Leadership.

Geschka, H. (1993). Visual confrontation—Developing ideas through pictures. In H. Geschka, S. Moger, & T. Rickards (Eds.), *The power of synergy—Proceedings of the fourth European conference on creativity and innovation* (pp. 151-157). Darmstadt, Germany: Geschka & Partner Unternehmensberatung.

Lindsey, E. H., Homes, V., & McCall, M. W., Jr. (1987). *Key events in executives' lives.* (Report No. 32). Greensboro, NC: Center for Creative Leadership.

London, P. (1989). *No more secondhand art: Awakening the artist within.* Boston, MA, and London, ENG: Shambhala.

McCall, M. W., Jr., Lombardo, M. M., & Morrison, A. M. (1988). *The lessons of experience: How successful executives develop on the job.* Lexington, MA: Lexington Books.

Merritt, S. (1994). *Ins and outs of imagery* (internal paper). Cambridge, MA: Polaroid Corporation.

Morrison, A. M., White, R. P., & Van Velsor, E. (1992). *Breaking the glass ceiling: Can women reach the top of America's largest corporations?* (2nd ed.). Reading, MA: Addison-Wesley.

Musselwhite, W. C., & De Ciantis, C. (1993). Learning to create shared vision. In S. Gryskiewicz (Ed.), *Discovering creativity: Proceedings of the 1992 international creativity and innovation networking conference* (pp. 91-95). Greensboro, NC: Center for Creative Leadership.

Parker, M. (1990). *Creating shared vision: The story of a pioneering approach to organizational revitalization.* Oak Park, IL: Dialog International.

Sayles, L. R. (1989). *Leadership: Managing in real organizations* (2nd ed.). New York: McGraw-Hill.

Shuman, S. (1989). *Source imagery: Releasing the power of your creativity.* New York: Doubleday.

Vaill, P. (1989). *Managing as a performing art: New ideas for a world of chaotic change.* San Francisco: Jossey-Bass.

Young, D. P. (1995, February 23). *Does training for action work: Key findings from the LeaderLab impact study* (Research Colloquium). Greensboro, NC: Center for Creative Leadership.

Young, D. P., & Bauer, A. (1993, May 13). *LeaderLab formative evaluation: Summary report* (internal document). Greensboro, NC: Center for Creative Leadership.

Young, D. P., & Dixon, N. M. (1995). Extending leadership development beyond the classroom: Looking at processes and outcomes. In E. F. Holton (Ed.), *Academy of Human Resource Development 1995 Conference Proceedings.* Austin, TX: Academy of Human Resource Development.

Young, D. P., & Hefferan, J. (1994, October 4). *LeaderLab program feedback report* (internal document). Greensboro, NC: Center for Creative Leadership.

Zaleznik, A. (1992). Managers and leaders: Are they different? *Harvard Business Review, 70*(2), 126-135.

Appendix A: LeaderLab Program Content and Structure

The stated purpose of the LeaderLab program is "to encourage and enable leaders to take more effective actions in their leadership situations." This embodies the idea that the leader of the future needs the resources to act, and not just reflect, in order to confront the serious challenges that await him or her. Based on this idea, the LeaderLab program content falls into four basic categories: (1) the significant generic challenges faced by leaders today; (2) the competencies necessary to deal with these, or any, challenges; and (3) the skills and knowledge that help each participant to understand his or her specific leadership situation and to take action. A fourth category, the information and experience that each participant generates by means of number (3), is developed in the course of the program and is unique for each individual (Burnside & Guthrie, 1992).

The content and structure of LeaderLab reflect principles that are appropriate for any training course but especially for one where action is emphasized. These principles are: *realism* (that teaching leadership has to take into account the complexity of factors that are part of every leadership action); *simplicity* (ensuring that program content does not overwhelm the participant); *relevance* (making content always relevant to the goal); *operationalism* (ensuring that content lends itself to being put into practice); *holism* (that training must be conducted in terms of the person's intellect, emotions, and behavior [referred to in LeaderLab as head, heart, and feet]); and *intervention over time* (ensuring lasting behavioral change). The last two in particular are critical to LeaderLab's approach. The importance of these principles resulted in a unique set of program features designed to support them.

The LeaderLab Model

Developed for the program, this model integrates four critical leadership challenges for the future with five competencies. The challenges, identified through surveying more than 100 leaders from Fortune 500 companies, are: dealing with rapid and substantive change; managing diversity of people and views; building the future through a shared sense of purpose; and the crucial challenges of each leader's particular situation.

The competencies, largely based on the work of Leonard Sayles (1989), "can be thought of as a comprehensive behavioral overview of the tasks of a leader" (Burnside & Guthrie, 1992, p. 10). They are: dealing effectively with interpersonal relationships; thinking and behaving in terms of systems;

approaching decision-making from the standpoint of trade-offs; thinking and acting with flexibility; and maintaining emotional balance by coping with disequilibrium. These competencies are distinct yet overlapping and intercon-nected. According to Burnside and Guthrie (p. 13), successfully meeting each of the challenges requires all five competencies.

Learning to Learn and a Sense of Purpose

Again and again in the program, from the precourse situation audit completed by each participant through follow-up with the process advisor on action planning, the participant is refocused on the reality of her or his own unique leadership situation. In this context, two elements which are crucial to how leaders develop themselves are also critical parts of the program's conceptual model: learning to learn and develop and sense of purpose. Center research has shown that successful leaders learn from experience (Lindsey, Homes, & McCall, 1987; McCall, Lombardo, & Morrison, 1988; Morrison, White, & Van Velsor, 1992). Moreover, the greatest potential learnings may come from situations "that one finds personally most difficult—that is, going against the grain of habitual practices" (Burnside & Guthrie, 1992, p. 14).

Facing today's leadership challenges, and finding the energy required to internalize the difficult lessons of experience are not easy. Strong motivation is required to persevere. Awareness of sense of purpose is therefore central to the LeaderLab model. Sense of purpose involves recognition of an individual's direction and internal focus, *and* identification of the unique requirements of his or her current leadership situation, or external focus.

Program Structure

Based on the challenges and competencies outlined above, different elements of instruction were combined to reflect this content and maximize results. The basic components of the program are outlined below.

Intervention over time. Intervention over time is a critical factor in the program's approach to ensuring lasting behavioral change (Burnside & Guthrie, 1992). The program is therefore structured to accommodate the full adult-learning cycle to occur (planning, doing, and reflecting). It begins with a six-day classroom session culminating in each participant's development of an action plan for improving his or her leadership. Following three interven-ing months back on the job during which the participant works to carry out the plan, a second classroom session of four days is held. At this time, partici-pants review how the action plan progressed, and modify it based on their experiences and learnings from the first three months. The revised action plan set at the end of the second classroom session is then implemented over the remaining three months. The entire intervention lasts six months, during which time participants receive the support of an individual process advisor who serves as a personal coach. Change partners within the participant group provide support and help to process learning. Work-site change partners provide feedback, support, and advice on behavioral, organizational, and technical goals.

Program activities in an action-oriented leadership program. The focus on learning to learn and relating learning to the individual participant's unique leadership situation called for classroom activities in addition to lecture. These include discussion, exercises, simulations, use of a learning journal, and nontraditional activities. Discussions promote cross-fertilization of ideas among participants, allowing them to bring their experience and expertise to each other. Exercises "serve as the first safe place that new ideas and behaviors can be put into practice" (Burnside & Guthrie, 1992, p. 22). The simulation is an extended exercise (known as "simmercize" in LeaderLab) which puts participants into the organizational roles of a fictional company, dealing with rapid change and difficult issues while wrestling with seemingly conflicting individual and divisional values. The daily learning journals provide a vehicle for reflection on learning and "provide the individual with themes or patterns of behavior that can be useful in taking more effective actions in one's situation" (p. 16). The nontraditional activities are critical from the holistic standpoint of conducting training in terms of the whole person: intellect, emotions, and behavior (p. 5).

Multiple sessions. LeaderLab begins with a six-day session where, among other things, an action plan is developed by each participant to improve his or her leadership, followed by a three-month interval during which the participant carries out the plan, and ending with a second four-day session, during which time the action plan is reviewed and another three-month interval is set up. Thus, the program duration is six months.

Process advisor. This is a staff person who meets with the participant during both weeks of training and helps construct his or her action plan. The advisor also contacts the participant monthly by phone during the three-month intervals. The advisor continually prods and encourages the participant to address issues and blocks in his or her development.

Change partners. The main task for each participant in this program is to create and follow through on an action plan for his or her leadership situation. A system of support, both in the program (change partners) and in the workplace (back-home change partners), is set up to support the participant in this goal. In LeaderLab, a diverse, three-person, in-course work group collaborates to encourage the individual. At home, each participant must also establish a group of change partners in the organization to help him or her with leadership improvement.

Diversity of participants and trainers. The LeaderLab classroom composition is a mix of gender, ethnic groups, and work situations, since diversity is one of the challenges of the future. A working actor and artist as trainers have added an important element of experiential diversity.

Traditional learning activities. These include instruments to assess strengths and developmental needs; lectures; discussion groups to promote cross-fertilization of ideas; exercises where participants can experience the theory being discussed; simulations (called "simmercizes") which are extended exercises which put participants into organizational roles of fictional companies; and the use of a daily "learning journal."

Nontraditional learning activities: art, acting, and sculpting. The artistic component, primarily the touchstone exercise, is discussed in depth above. A number of additional components which have not been traditionally part of a leadership development program are presented in LeaderLab.

The acting component was designed and is trained by a working actor and incorporates methods used in acting instruction. Its focus is on awareness of participants' physical self-presentation (tone of voice, body posture, gesture) and its effect on others. The "Acting Leader" is the first of the nontraditional segments to be presented in the program. It comprises two classroom sessions in week one of the program, plus a personal assessment provided by the trainer of each participant's behaviors as observed during the opening "simmercize." In a brief individual session, each participant is given a "suggesture" (from suggested gesture), based on what he or she typically tends to do, that would make the participant more effective (Burnside & Guthrie, 1992, p. 23). Two more sessions take place in week two which focus more on group interaction based on the same principles.

The body-sculpting component was designed by a practicing family therapist and makes use of the pioneering work of Virginia Satir in identifying and working with family dynamics issues non-verbally. In week one, this session occurs on the second day. In its adaptation to the leadership development context, participants work non-verbally with an issue or situation that has caused difficulties at work, using classmates to stage a scenario, representing first a problem situation, then creating a scenario representing a possible solution. The scenarios are verbally debriefed in depth. A second session occurs during week two, during which participants form groups to create scenarios representing problems they have faced in meeting their action plans, and then creating a scenario representing resolution of these difficulties. Because the participants use their bodies as "clay" to "sculpt" their own unscripted scenarios, these exercises are relatively unstructured. They require skilled facilitation, both in the process and in the debrief which involves being able to translate a non-verbal experience into terms relevant to the participants' work environment.

Appendix B: Conducting the Touchstone Exercise

The information contained in this appendix is a description of how the Center conducts the touchstone exercise. This is not meant as a guideline. The reader is encouraged to use whatever pieces of it that he or she finds useful.

Introducing the Exercise

Before the exercise begins, participants should be introduced to the notion of the non-verbal "excursion" and its value in circumventing our accustomed verbal logic to give a new view of the subject of inquiry at hand (e.g., the participants' leadership issues or focus). Providing this kind of orientation is particularly important if no other nontraditional or non-verbal exercises have been presented previously in the training context. Request participants to suspend judgment while they participate. Ask for questions to make sure participants understand the purpose of the exercise. For the exercise to be most effective, appropriate and clear links should be made with other program content, both before the exercise and during the debrief.

Since the touchstone becomes a representation of something personal, it is helpful, although perhaps not necessary, for the trainer to provide the participants with a personal example. If an artist facilitates the exercise, this could involve the sharing of one of the artist's works and telling a story about it which is at the same time personal to the artist as well as accessible to the audience's experience and related to the overall theme and goals of the program context. A non-artist might introduce the exercise with a personal anecdote related to the program's themes (for example, a personal experience involving an unusual learning method and what was gained from it).

Most important is framing the question, or focus, which the participant will be answering for him or herself in the creation of his or her touchstone. The participants need to know what it is they are being asked to represent. At the same time, the question should be open-ended rather than specific and detailed. For example: "Think about what is most important to you in your leadership. What is it you most want to be able to remind yourself of when you are back in your individual leadership situation?" While open-ended, the focus should be central to the program context's objectives.

In the specific context of the LeaderLab program, participants' attention is redirected to the center of the LeaderLab model, sense of purpose. They are invited to give concrete expression to their own sense of purpose, "to YOU, and what you uniquely bring to your leadership, the sum of all your history, your experiences, your values, what's most important to you."

Participants are then given details about the process. They are told they will be provided with materials to build their touchstone, and to be aware of colors, textures, shapes, sizes, even smells. They are asked to be thinking about what is most important to them in their leadership, but "don't think too hard, just pick up anything that speaks to you about what is most important to you in your leadership." They are asked when they have finished their touchstone to sit with it and "let it tell you its story," and to record that story in their journals. A written record of the story is an important part of the exercise. The intent is not to abandon words; rather to circumvent verbal logic for a while, but then to return to words. When the group reconvenes, they will be invited to share stories publicly, if they wish. And, at the end of the program they will take their touchstone with them (or it can be shipped). They are asked to put it in a place where they will be able to see it. This can be anywhere they like—desk, living room, closet—as long as they will periodically look at it and let it remind them of what is most important to them, and what they most want to remember from the program experience. Having them take it home and revisit it is key to the usefulness of this exercise in transfer of learning.

Before leading the group to the exercise space (which can be in the training room or in a separate room), I tell them the expected time that the group will reconvene following the exercise. Forty-five minutes is the typical amount of time given to the exercise in management groups.

Facilitating the Exercise

Creating a safe and non-judgmental environment is key to the success of this exercise and requires a balance between lightheartedness and the expectation of a depthful experience. Although appropriate humor is a useful bridge in almost any situation, I try at the same time to avoid making light of the exercise or suggesting anything of the kindergarten room. The exercise is both fun and serious and worthy of attentive and earnest participation by adults. The demeanor of the facilitator should reflect this and that the exercise will be carried out in a non-judgmental atmosphere of encouragement.

The product is ultimately less important than the process, although respect should be shown to the products without dishonesty or overpraising. The touchstones produced by participants should be expected to vary in artistic quality, but it should be understood that all are equally important to their creators. It is important for the facilitator to show respect to both the process and the participants. It is also important to keep in mind that the touchstone exercise is not intended to be diagnostic (although it certainly could be used as a diagnostic tool under other circumstances). It is wholly and simply a vehicle of self-reflection and its value should be understood to be unique to each participant and honored as such.

As participants complete their touchstones, I remind them individually to sit with their touchstone for a few minutes and "let it tell you its story," then record that story in writing. I also remind them that the group will be reconvening afterwards to share stories. In my experience with this exercise, participating managers and executives have not objected to or seemed to have undue difficulty with such language as "anything that speaks to you" or "let it tell you its story." Suzanne Merritt, the originator of a similar exercise, "Visual Voice" (described in Appendix D, "Examples of Artistic Methods Used in Organizations") uses similar language with her organizational constituency. I usually ask each participant to tell me what their touchstone means before they leave the room. Most are willing to share the story, usually with enthusiasm. During the exercise period, the needs of participants who are still busy building come first and it may not be possible to personally ask for each and every story. Read the section titled "Resistance to Artistic Activities" above for profiles of some participants who had a difficult time with the exercise.

With regard to technical facilitation, some participants may be hesitant or shy at first when confronted with the exercise materials. I introduce them to the materials by saying "everything you see is yours. Take anything you want, and as much of it as you want. Take anything that speaks to you about what is most important to you in your leadership [or whatever is relevant to the open-ended question at hand]." Most people need little direct facilitation in terms of what or how to select objects. Some will require a little gentle prompting in the form of open-ended questions or a reminder to simply pick up anything that speaks to them about their leadership situation [or whatever focus they have received]. As people start to pick things up, I tell the group that, as they start to put things together, to "be aware of the meaning of each thing and the relationships of the things to the whole." This is as much an aesthetic as a facilitative prompt. Much of the meaning of the touchstone will reside in the relationships of elements to each other, and that same relationship also represents the most basic building block of artistic structural values: composition. Few, if any, other instructions seem to be necessary to get participants started with the exercise.

When making their touchstones, some participants will need a bit of technical assistance: what glue to use, how to put certain things together in the way they want them. Often, they will assist each other. Sometimes, a participant may wish to do the impossible, for example to stand a large or relatively heavy object on end. I will often suggest that the participant find an additional object or objects to prop it up. Such an arrangement is usually seen by the participant to be significant to the overall meaning of their touchstone and will frequently itself show up as an element of the touchstone's "story."

A few technical details are given below in the section headed "Construction materials and tools."

Debriefing the Exercise: Sharing the Story

Participants are reconvened as a group following the exercise to share stories upon completion of the exercise. Along with the actual creation of the touchstone and its use as a concrete reminder of sense of purpose or "what's important," the public sharing of the stories comprises a large portion of the power and relevance of the exercise, and is felt by the participants to add significantly to the value of the experience. Safety is important here. No participant should be made to feel pressured by the facilitator into sharing his or her story publicly.

If participants are straggling in at the last moment, it can be helpful to start by inviting participants to get up and tour the classroom to get a look at the "gallery" of touchstones. They will see that from the same pool of materials they have individually created very different works. A tour of the "gallery" tends to be enthusiastic and can become boisterous with sharing of individual stories and (mostly) good-natured teasing.

Once participants have reconvened and are seated, I simply ask: "Is anyone willing to share their touchstone story?" Silence on the part of the facilitator following the repeated request: "Anyone else?" is usually sufficient persuasion. Peer pressure often comes into play. This is usually good-natured, although in some instances it may cause one or more participants to feel unavoidable pressure. Avoid prompting participants by name and concentrate on maintaining a safe, friendly, and inviting atmosphere. Simply provide space and silence for the next speaker, and thank each for their story.

Practical Considerations

Physical space. In situations similar to that described in this paper, it is desirable to use a room separate from the training room for participants to work in while creating their touchstones. This allows for materials to be set up without distracting from training activities which precede and follow the exercise.

Enough space is required in order to provide tabletop working room for each participant, plus a separate table for materials. Chairs may be provided around worktables if desired. Access to electrical outlets is required for the use of hot glue guns. In a typical program group of 21 participants, three to four 3' x 5' tables are placed against the wall to provide working spaces, and two or three tables of similar size are placed together in the center of the room for the display of materials.

Given enough time, an option would be to conduct the exercise out of doors, each participant selecting some or all materials during a stroll through the natural environment.

Exercise preparations. In the exercise room a large quantity and wide variety of natural and man-made materials are laid out attractively on a large table. A list of suggested materials is provided below.

The materials can be laid out randomly on the table, to suggest the profusion and confusion of nature. Recently, when working with a group of mostly engineers, I laid the materials out with a neat and ordered directional orientation to reduce the potential confusion and distress of their being confronted with initial disorder. The subtle colors and forms, especially of natural materials, can sometimes be made more visible to participants by the way in which materials are presented. For example, materials which are the same color or a similar tone can be placed together.

Worktables should be covered with brown craft paper or newspaper for protection from glue. Tools and glue are distributed between the worktables. A list of suggested tools and supplies is provided below. Glue guns should be plugged in and take a few minutes to heat up.

Suggested materials. As large a variety as possible of textures, shapes, and colors, even odors, is important. Providing a large quantity of natural materials has the benefit of being free of cost. It also gives an impression of abundance which lessens for participants any potential anxiety when selecting their materials. The more that is provided to them, the less they will feel there is a "right" thing to select.

> Bases: cut-out pieces of one-quarter inch plywood of varying sizes and shapes. Other foundation materials can include weathered wood, chunks of bark, and large rocks.
>
> Dried leaves, twigs, and small branches in varying lengths. Other natural materials can include dried flowers, pinecones, bark, moss, thorn branches, "tree ears" (dried fungus), sea grass, acorns, seeds, and pods. Fragile materials such as leaf lace can be used. Try to find rocks in a variety of colors, sizes, and textures from

smooth to rough. Feel free to include materials which are unique to the geographical area.

Seashells and other beach materials. If you gather natural materials from the beach, be sensitive to unpleasant odors created by some of these materials as they dry. Seashells are available bagged from craft stores.

Polished colored stones such as variegated agates, tiger eyes, amethyst, rose quartz, coal, hematite, apache tears obsidian, malachite, clear quartz crystal, and native copper. Polished stones provide color, hard textures, and luminosity. These can be purchased from natural science centers, toy stores, and stores which sell natural items as gifts. Hardness and color are also provided by marble-sized, flat-bottomed glass "gems," available inexpensively from stained glass suppliers.

Soft materials: fabric of varying colors, weight, and textures; colored paper; mylar wrapping paper; patterned fabric scraps; fur; feathers.

Miscellaneous objects: aluminum or copper sheeting; rusted metal; broken colored glass and mirror; weathered rubber; string or twine. Polymer-coated outdoor clothesline wire in different colors is available from hardware stores, as is copper, brass, and metal wire.

Participants occasionally request paint. School-grade tempera paints can be provided in jars or plastic cups along with inexpensive brushes, paper towels, and cups of water for mixing. Providing a separate brush for each color is useful for participants who have no recent experience with paint.

Construction materials and tools.

Masking tape (2-4 rolls). This is useful for securing materials in the desired position until glue sets. Other objects may also be used to support materials in the desired position (for example in unusual angles or vertical positions) until glue dries.

Low-odor silicon caulking and caulk guns, available in hardware stores. This is useful for non-porous materials and may hold stones and some heavy materials more effectively, as well as heavier materials the participant wishes to stand upright or at a vertical angle. It should be applied generously. It does not set rapidly, requiring about half an hour or more if copiously applied, and 24 hours to set thoroughly. The bond, though permanent, may feel flexible even when set.

Hot glue guns and heavy-duty glue sticks, available in hardware and craft stores. Hot glue is easy to use and sets rapidly. Used copiously, it may be used to secure lightweight materials in an upright, unsupported position. It may not hold smooth stones or other non-porous materials permanently. When used copiously, it takes a few minutes to set up. Warn participants to avoid getting the hot glue on their fingers.

Optional glues: white glue, gel-type "crazy glue," glue sticks.

Tools: scissors, x-acto knife, wire cutters, tin snips, glass cutters, and breaking pliers (for use with glass and stained glass, available from stained-glass suppliers). Participants often ask for a drill which I have never supplied, but it is an option.

Packing materials. If touchstones are to be shipped to participants, the following materials are required:

Cardboard packing boxes (12"x12"x6" is a typical size)
Packing tape
Bubble wrap
Newspaper, shredded paper or shipping "peanuts"
Labels

Packing. When packing touchstones, remember they are works of art. Wrap touchstones in bubble wrap, being careful to protect fragile or protruding parts. Use newspaper, shredded paper, or "peanuts" to surround the piece firmly but gently and fill the box completely to provide maximum cushioning. Extra wrapping is necessary for excessively fragile pieces as well as very heavy pieces, which might damage themselves by shifting during shipment. Surround fragile pieces less firmly with newspaper, shredded paper, or "peanuts" to avoid damage. With very heavy pieces, be sure the object is well-cushioned and firmly secured within the box so it cannot shift.

Appendix C: Touchstone Stories—A Sample from One Group

What follows are the stories told by most of a single LeaderLab partici-pant group at the end of the second week of the program. Four participants were female and one was African American. Their organizations included a not-for-profit health advocacy group, federal and state government agencies, financial, chemical, manufacturing, and information technology companies. These stories have been transcribed verbatim from audiotape, with permission from the storytellers.

In the second week, touchstone-exercise participants are asked to represent their work group in its shared sense of purpose. Some make refer-ence to the metaphor-survey responses returned by members of their work groups.

Story 1. What I have here is a touchstone to represent the group that I work for, and when I had them turn in the metaphors for our work group they described what we had to do as big, as difficult, and having frequent changes. So the concept of this piece was: big, hard, and frequent changes. Now, there are ten of these green squiggly things and I thought that they were more flexible than they really were, so that was supposed to demonstrate flexibility, but it's also a fact in the real world that people aren't as flexible as you think they are. They also go off in all different directions. My team is always traveling around the world, so occasionally we're in the same general area but not very frequently, and what those seven colored stones represent is in order to sell the vision, you have to sort of market it, so that's a little bit of flash, but it's more than flash, it's a holistic approach because what those seven stones represent is the McKinsey 7-S Model. If you're not familiar with that, it sort of says that in order to institutionalize your change, you have to deal with seven rather large categories: you have to deal with strategy, you have to deal with structure, you have to deal with shared values, you have to deal with skills, you have to deal with systems, you have to deal with style, and you have to deal with staff, so that's what the seven colored stones are aside from the marketing. So that represents my work group and what my vision is, and to refresh your memory, our vision is to make [company name] recognized worldwide as a premier software company, or a company that has software core competency, by 1998. Any questions?

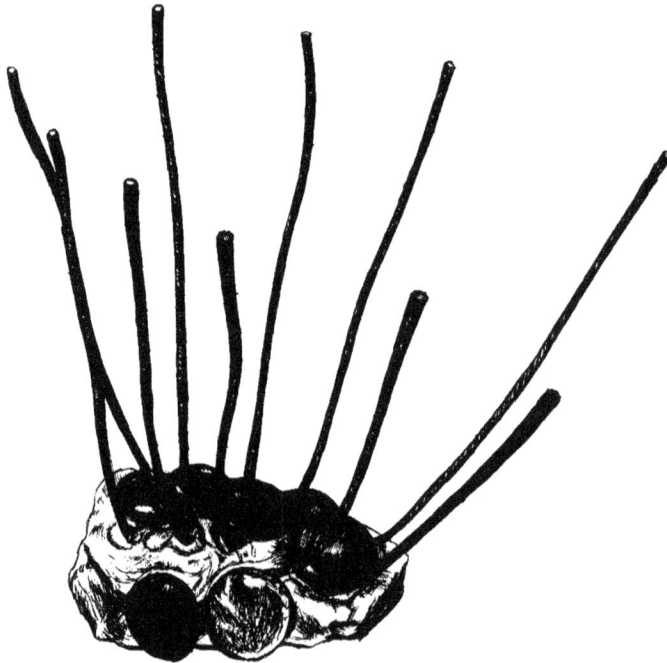

Story 2. The themes and the metaphors that people gave to me or for me of me about me (prepositions have always been difficult) was about—had to do with the role of influencing in a larger system. What I've tried to represent here is the future in the center is represented by seeds and each of these figures along the perimeter are individuals viewing the future. And at their core these individuals have something in common: these pieces of quartz, which is somewhat clouded by the fissures and fractures in them, but as they look towards the future from where they're standing, they are filtering that view of the future through their experiences of the past. So what this represents to me is to recognize and accept that people are going to have those differences and to work with them to discover our common similarities and the clarity that result from that.

Story 3. Helps if you grew up building models. Essentially my vision as it's still being worked on—but I'm using the metaphor of the space program of the sixties as the way I'm encompassing my vision, so I'm using a lot of words like "mission" and "challenge" and "pushing the envelope" and those types of things as part of my vision. So what I really wanted was something that looked fairly literal to something space and rockets and things like that, but also this ties in fairly well with the metaphors that I got from my people because it also reflects a lot of the "height"—I got things in the sky and mountains, things that have a lot of height to them, and also a flock of geese and that's why this is kind of a delta shape to kind of tie into that idea of a flock of geese.

Story 4. I think that I sometimes make life too complicated. So, I chose something very simple. Basically, the vision is—I think that my responsibility for the vision is to provide the frame and support and I chose shiny materials because the vision has to be a reflection of the people who work in the organization.

Story 5. I wanted to use my touchstone to stab anybody who got out of line in the work group! [laughter] This actually represents the work group that I am just coming into. This is kind of an operating table and on the center of the table is supposed to be an infant, and the work group is supposed to be the infant. And to me an infant represents such a miraculous view of the future, because the growth is unbelievable, the learning is unbelievable, and so I see the work group as an infant right now. Contributing to the infant is a sense of preparedness, and that's what this book is right here, and the idea is that for the infant to grow, preparation above and beyond what most people would do will be required. Use of technology and communications, which is what the Wall Street-like headset is—and that's supposed to depict the communications style and pace of the environment. And the last little heartwarming thing over here is a scalpel that's supposed to go along with the surgical table. And it represents a requirement of executing with surgical precision.

Story 6. I don't know whether I've got a metaphor for it, so I've built it anyway. Basically what I've tried to do is to provide a backdrop of an environment which has a lot of rough edges, which is this piece around here, but have essentially a collection or a team that works together that has some form of communications or boundaries. I purposely put two rather interesting circles around: one very rigid and very geometric, because that I think is sometimes some of the framework of the organization, but in reality most organizations don't work that way, so what I've done is tried to put some form of informal communication on there with the idea being that the product is actually the focus of attention, which is the center part, but I purposely put that down at the bottom because I've really for me believe that when it comes down to it the products of the company or the output of the company is important, but it's not necessarily *the* most important thing. I believe the people and the structure that goes toward it is actually more important than the end result. That's my naive approach to [personal...] and I'm sticking to it!

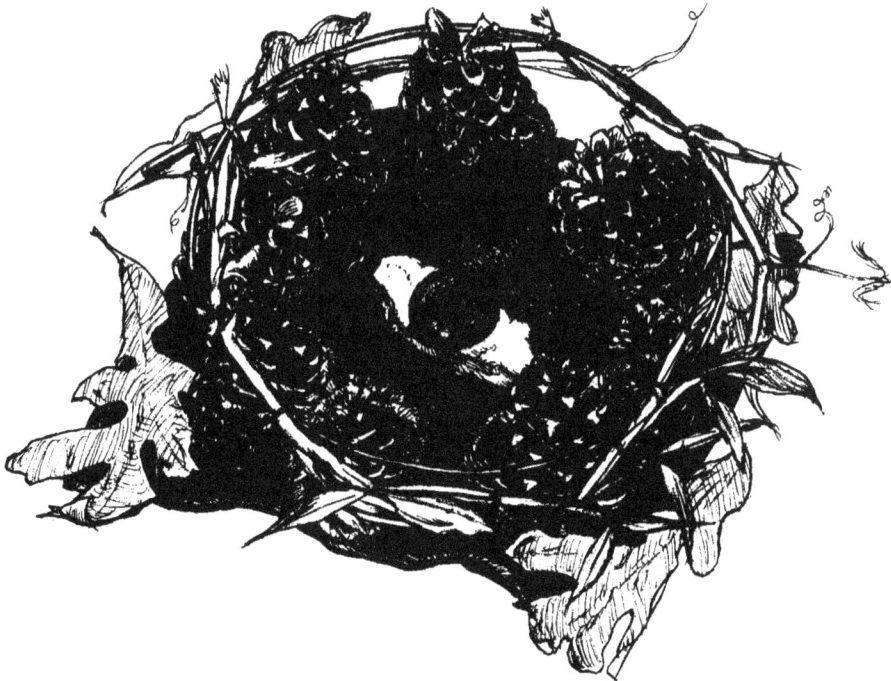

Story 7. I started out doing something similar to what Jim did and I couldn't figure out how to do it and I guess on the spur of the moment I came up with this. If you were to turn this and look at it from the top, for most of you that have done math and physics, it looks like the symbol of infinity. Basically change goes on forever. I had about six or seven rings and I kept cutting back with really no rhyme or reason and I ended up with four, which is like the four dimensions—the inner core is also disconnected and represents change. That is the core of our work group and that work group changes as well: people retire and people coming in and such. And it's that core that manages the change and keeps it all together in some manageable form. The changes themselves are not connected, so when these things—they'll slide back and forth so one of these can get bigger and another one gets smaller, which means that the changes aren't all the same: there are some smaller changes and some big changes. That's about it.

Story 8. I shall demonstrate. This is—this little thing here is 1994, and this is 1997 down here. And this is what I want for our division which is excellence, value and leadership, which will be the cornerstones on which we'll build the future. And this is our customers, our broader organization, our employees and our selves, meaning each person who works on this. And each of these is the viewpoint that each constituency has regarding our organization for each of these key areas, all pointing towards where we want to be.

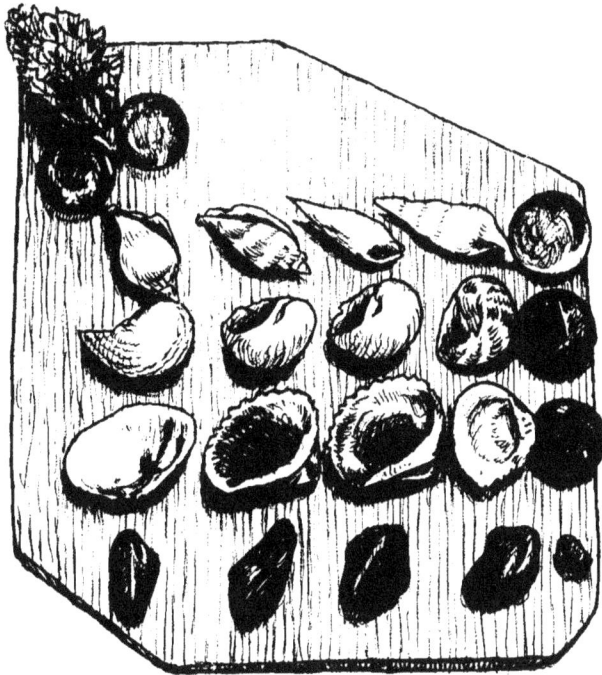

Story 9. This is life! In our business or what we're trying to do—but it does have some overtones in just general things that you work yourself up a ramp: it's fairly easy, predictable, you can work your way up. As you get there you get to a point where you have to make some decisions and the direction you're going to go. In our business some of the people will basically soar like a bird and take off. Others are going to be very rooted, very strong, very stable, provide the support for these others. But as you go along occasionally you'll find these rocks in the road or problems that come up, and we've had these kinds of problems. We've had abuse problems, signified by the wine cork, so on occasion you will have a cloud come in front of the sunshine, but eventually that clears out of there. We've had people that become very bright, shiny and blossom, and that cloud comes back and we've had others that wither and fall into ill health and occasionally leave us, but these rocks continue. As you see we occasionally run into ogres, but for the most part, we find that people are gems. Sometimes all of us determine that the grass is a little greener somewhere else and you have decisions to make as you then come against another wall, and as you climb that wall and scale it and feel comfortable you find that your choices become a bit easier and you can coast toward the next phase of life, and eventually there will be a recycling.

Story 10. What I'm trying to represent—I had a lot of thoughts going on in my mind so unlike [participant no. 4] who really went for simplicity, I'm still striving for that clarity for my work group. I have a really high-performing team right now, but I really want to continue that momentum and to create an interdependency between the three groups, so that's what the pinecones represent. And I want to provide a nourishing environment for them. I've taken objects from nature to display that: growing, constantly changing, constantly having new growth, although recognizing that there is some chaos in that. The copper wire is basically my leadership providing energy, conductivity of ideas, feeling, sensitivity; and the quartz prism is actually something somebody had said our group represents in the metaphor as a constantly changing, dynamic environment open to new ideas, new relationships, new ratios, but that we're sort of always floating along the stream and flexible, a springboard of ideas. Lots of thoughts there.

Story 11. I had a hard time figuring out what my work group was since we're in such chaos, and I just decided that I had to look at all [company name] in terms of how to put my vision together. And I thought that what we need to succeed to become a profit-oriented company with these major changes is we need professionalism towards our three major markets and our customer interface and sales organization. We need to be creative and that's what the metal's supposed to be: creative in our approach to our solutions to our customer and this is supposed to be our resources, and we need to look through it with business acumen, get some clarity as to how we look at those resources and the creativity, which we haven't been real good at doing at [company name]. If we do that, then [company name] will thrive. Right now [company name] is divided into three areas: a marketing area, a product area and a production area. So for [company name] to survive actually we have to get that stuff [on a roll? through?]

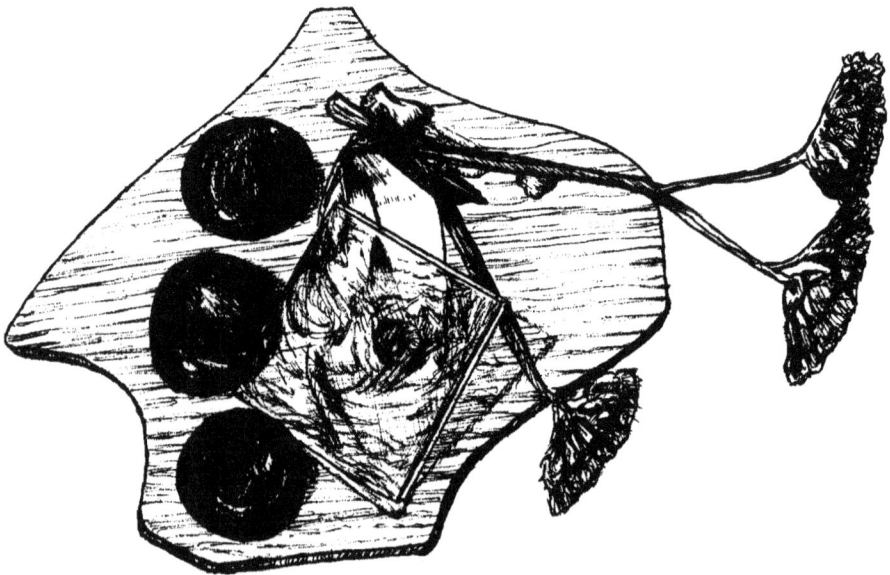

Story 12. Still thinking about transition. The left side is the old; it's either a dragon breathing fire or Viking ships or whatever you happen to see in it. Whatever it is, it's constrained and it's old and it won't live much longer. We're trying to get into a new vision but as part of education you're the bridge between the old and the new and I find that the new vision isn't really there; in fact it shouldn't be quite this, it should be twisted 90 degrees and it should be very calm, and it should be a blue reservoir. From this data people can get all the information they need. It's a little smaller than they think it needs to be, but I haven't quite figured out the way to get from the old to the new. I'm going to have to figure that out in terms of the training, probably before the real vision is complete.

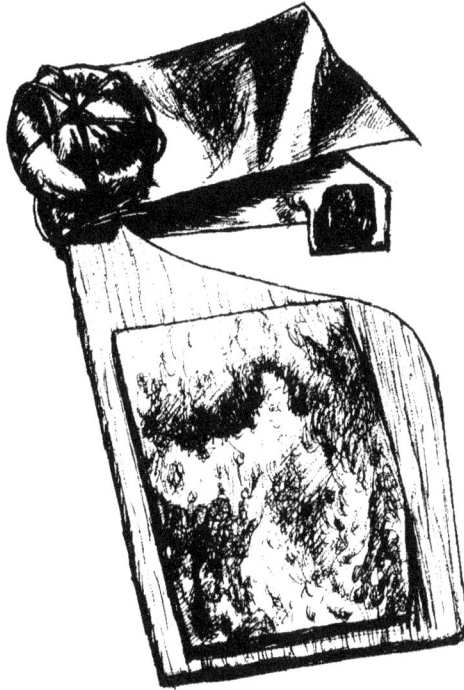

Story 13. Basically what mine is is this is the new organization we're currently trying to form: our separate division made up of four operating units. You can't see it, but each unit is going to be in its own little shell but loosely connected with some copper wire. Now what we have to do is to figure out how to get into some protection out of this harsh area. So we're new and shiny, which road do we take or which path and that's what my work group is supposed to be working on is getting the structure and direction we need to go.

Story 14. The vision I would have had if I was able to express it better would have been an octagonal globe with a spacecraft going around it. Then there'd be a force that you'd be able to see on the globe. Let me describe what this is: the octagon is [company name]'s corporate symbol, it's been there for 90 years. The octagonal globe is '#1 in the world.' That's our goal to end up in control and be number one in the world. The forest represents the products that my group builds, so I would have a forest coming out of it. The spacecraft is the organization kind of going through a harsh environment of the rest of space, and people stay protected and nurtured inside, so if I had the tools to do that I would express that.

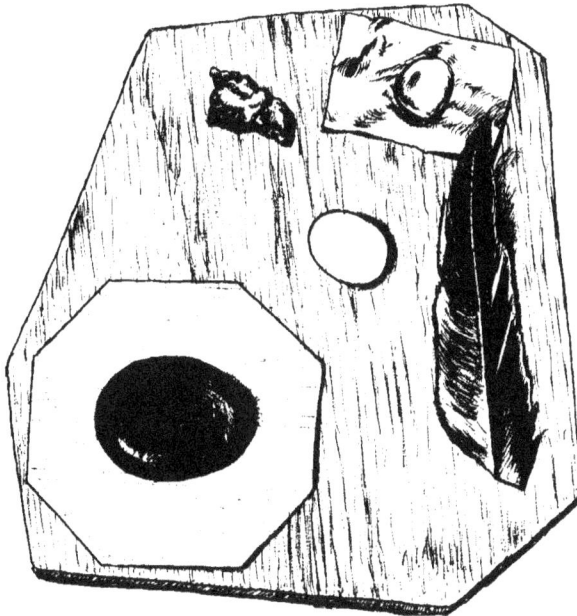

When participants were asked to indicate by a show of hands which of them would actually be likely to share the story of their touchstone with their back-home work group, approximately three-fourths of them raised hands. One said, "I'm going to put it in the middle of the table and ask my work group to give me an interpretation before I share it with them."

Appendix D: Examples of Artistic Methods Used in Organizations

From 1986 through 1988, Marjorie Parker of the Norwegian Center for Leadership Development conducted an organization-wide visioning project in Hydro Aluminum Karmøy Fabrikker, described in her book *Creating Shared Vision* (1990). This project provided participants (comprising the entire organization, working in small cross-level, cross-functional groups over a period of eighteen months) with the opportunity to employ multiple techniques including songs, poems, limericks, sketches, collages, models, and drawings to explore, develop, and ultimately own a vision of KF as a garden, a metaphor which originated with the managing director of the facility.

The aim of this project was to "lift the company to a new plateau" following a successful decade of turnaround from a crisis state of polluting and non-profitability to profitability and a commitment to a cleaner environment and employee development. The desired state past this plateau was a "revitalization of the organization which could lead to shifts in attitudes and beliefs and create a basis for continual self-renewal." The garden vision in its final pictorial version, rendered by an artist at the end of the process (an artist was not used as a facilitator) captured the seeming paradoxical fundamental organizational needs of "being 'apart' *and* being 'a part of'; of being independent *and* dependent; of being responsible for oneself *and* the whole; and of the whole being greater than the sum of its individual parts." Reporting on KF two years after the vision was created and resulting action plans put into effect, Parker states that "record breaking has become a part of everyday KF life. Higher productivity, lower energy consumption, and higher electricity yield are a few examples of where the records are broken almost every month. . . . Productivity has increased 33% from 1986 to 1989." Parker also reported positive impact on accident rates, employee absenteeism, employee health, and employee ideas for innovation resulting in benefits to the company amounting to millions of kroner.

In 1992 Suzanne Merritt, senior creatologist at the Polaroid Corporation, founded the Polaroid Creativity Lab (Merritt, 1994). In her previous role as a worldwide strategic planner for the entertainment-imaging segment of the organization, she saw limits in the corporation's ability to imagine its future, resulting in blind spots to potential growth areas. Merritt developed a five-year strategic and tactical plan for supporting creativity throughout the organization. She gained approval for her plan directly from the CEO and corporate officers as part of a total quality initiative which views creativity as contributing to success.

Since its opening, more than one thousand individuals and teams have attended sessions where they use a variety of techniques to work on real business opportunities. Merritt started with a number of methodologies current in creativity and innovation literature and practice. Prompted by the responses, enthusiasms, and spontaneous experiments she noted in her participants, she began to modify her basic methodologies and to add image-based techniques to spur the creativity of her organizational constituents.

Through practice, Merritt has identified the techniques most appropriate to the different stages of idea formation, development, and communication, as well as to types of problems. One exercise Merritt developed is called "Visual Voice." Inspired by the work of Sandra Shuman (1989), this exercise involves the creation of a collage from a pool of provided visual images (color photos or magazine pages) and can incorporate other elements such as paint, feathers, and glitter. According to Merritt, "Collage can be used if you are working on a problem which is abstract, complex, or has many different dimensions. These methods help get at the essence, core, or gestalt of your thinking to create a cohesive image where you can consider the many elements of your problem or idea at one time." Although participants are usually working on a specific business-related concept or problem which requires a creative solution, the exercise itself is similar to the touchstone in helping to "get at the essence." Merritt asks participants to form a mental impression of the concept they are working on, then put it aside, forget about it for a moment. They are invited to select images which appeal to them—"speak to them" as Merritt prompts—for any reason and to abandon themselves to the activity or creating something out of them. Merritt also provides toys and other objects in her laboratory; "the use of materials to make three-dimensional representation of [participants'] thinking worked well for ideas in development. This technique is especially good for establishing spatial relationships between different dimensions or aspects of a problem or idea and for constructing a dialogue around it." She describes one session when she asked a team to describe their organizational structure. Someone "jumped up and grabbed some blocks and materials and said, 'We are over here and they are over there.' He was defining their structure in a spatial form utilizing the materials at hand. The whole group got involved and after 45 minutes they had constructed a model of their structure which spanned the entire floor of the Lab. . . . For the first time, they each understood the relationships and interconnections between their various functions—something they had never been able to 'see' before."

Merritt began to see a link between aesthetics—the study of beauty—and organizational creativity. Merritt believes that "the leaders of the future will need to have highly developed capabilities in areas such as intuition, vision, holistic thinking, dealing with ambiguity, and most importantly powers of perception and imagination." She envisions an aesthetic education curriculum for executives that will "wake them up to the world around them and allow them to see things in a new way. Such a program would introduce managers to the importance of engaging their senses, intellect, and imagination, in order to perceive real needs in the world and create the means to meet those needs." In realizing this vision, Merritt expects to see resistance on both fronts, business and the arts.

> Individuals in the art and aesthetic field may not be interested in this relationship which may seem beneath them and not appeal to their principles of art for art's sake. Conversely, business leaders may find it difficult to believe that "learning to look" or "engaging the senses" could in any way be relevant to their work. However, I believe there is a great deal to be gained by creating new connections between these two worlds. Arts and aesthetics gain further validation of their contribution to society as a vehicle for rekindling the human imagination, and business becomes a place where the whole individual comes to work, with their senses and imagination intact, not just their intellect. (Merritt, 1994)

Merritt was invited to the Center by Bob Burnside to collaborate in the design of a new program, Leading Creatively. Teaming with program manager David Horth, Charles Palus, Pam Mayer, and Victoria Guthrie, Merritt has been working on the development of a set of "aesthetic competencies" for leadership. These competencies include perception (engaging the five senses as well as the extra "sensibilities" of feeling, intuition, and imagination), integration (recognizing the patterns and connections in often apparently disparate phenomena and fields of inquiry), engaging the preconscious (consciously accessing preconscious processes, such as dreaming), and visual thinking (the ability to apprehend and communicate data visually and metaphorically). This program offers exploration of artistic modes of inquiry, for example drawing and insightful writing. The aim of the program is to enhance participants' leadership effectiveness by enabling them to complement their already well-developed capacities for rational thinking with innate capacities for visual, holistic, and spatial thinking. The development of these capacities into competencies can enhance leaders' ability to address complex issues.

In another vein and also at the Center, San Diego artist facilitator Eva Montville collaborated with program designer Carole Leland to develop an artmaking exercise using collage for The Women's Leadership Program. This program acknowledges the need for women in leadership to be able to maintain a dynamic equilibrium between the demands of multiple roles they must play. The exercise is focused on the expression of the sense of personal balance these women are seeking, and out of which they will derive the energy to continually revitalize themselves in the face of multileveled, intensive, and continual demands in life and work. The exercise follows a half-day of working on an accounting of the multiple roles each participant plays in her professional and personal life and examining what elements of each provide her with energy and which drain energy. Each participant creates a collage drawing from a widely varied collection of provided materials. The participants are given ninety minutes in which to create their collage, and many use all of the time allotted. The collage is created within an acrylic box which becomes a dimensional frame which the participant takes with her when she leaves the program. The collage stories are shared with the participant group and can be an opportunity to open up as deeply as they choose in a safe and non-judgmental environment. One participant who chose to create a literal representation of a recent life event she felt to be transformative said: "It may look like something I made in kindergarten, but I'm just as proud of it now as I would have been then."

Ordering Information

To get more information, to order other CCL Press publications, or to find out about bulk-order discounts, please contact us by phone at 336-545-2810 or visit our online bookstore at **www.ccl.org/publications**.

www.ingramcontent.com/pod-product-compliance
Lightning Source LLC
Chambersburg PA
CBHW051420200326
41520CB00023B/7315